1983

Douglas M. Klieger

Villanova University

Computer Usage
for Social Scientists

Allyn and Bacon, Inc.

Boston London Sydney Toronto

For Cheri

Preparation Buyer: Linda Cox

Library of Congress Cataloging in Publication Data
Klieger, Douglas M.
  Computer usage for social scientists.
  Includes index.
  1. Social sciences—Statistical methods—
Data processing. 2. Statistics—Data processing.
3. Social sciences—Data processing. I. Title.
HA32.K58 1984   300'.28'54   83-2737
ISBN 0-205-07962-8

Printed in the United States of America
10 9 8 7 6 5 4 3 2 1   88 87 86 85 84 83

Contents

Preface   xi

=== PART ONE ===

1   INTRODUCTION   3
         1.1 Computer Availability and Use   6
         1.2 Time/Benefit Ratios   7
         1.3 Analysis Package Manuals   7
         1.4 How to Use This Book   8

2   COMPUTERSPEAK   13
         2.1 Hardware   13
         2.2 Software   16
         2.3 Codes   17
         2.4 Acronyms   18
         2.5 Old Words–New Meanings   19
         2.6 Structure   21

3   DATA COLLECTION AND PREPARATION   23
         3.1 Standard Data Format–The Matrix   25
         3.2 Data Collection   28
         3.3 Data Preparation   36

4   TERMINALS   43
         4.1 Types of Terminals   44
         4.2 Operating a Terminal   47

4.2.1 Keyboards   48
4.2.2 Summary   51
4.3 Operating Systems   52
4.4 Living with a Terminal   53
4.5 Special Problems   54

5   KEYPUNCHES   57
5.1 IBM 026 and IBM 029 Keypunches   58
5.2 General Operation and Mechanical Aspects   58
      5.2.1 IBM 029 Keypunch–Physical Layout
            and Operation   58
      5.2.2 Control Switches   61
5.3 Keyboard Details   62
      5.3.1 The Alphabetic Keys   62
      5.3.2 The Numeric Keys   63
      5.3.3 The Special-Character Keys   64
      5.3.4 Special Problems with Parentheses, Commas,
            and Quotation Marks   64
      5.3.5 The Space Bar   65
5.4 Keypunch Control and Control Keys   65
      5.4.1 Movement Control Keys: REL (Release), FEED,
            and REG (Register)   65
      5.4.2 Punch Control Keys: NUMERIC, ERROR RESET,
            MULT PCH (Multiple Punch), and DUP
            (Duplicate)   66
      5.4.3 Production Keys: PROG ONE (Program One),
            PROG TWO (Program Two), LEFT ZERO,
            MC (Master Card), AUX DUP (Auxiliary
            Duplication), ALPHA (Alphabetic), and SKIP   67
5.5 Automatic Control–The Program Drum   68
      5.5.1 Mechanical Aspects   68
      5.5.2 Program Card Operation Codes   70
5.6 Keypunch Automation Applied   74
5.7 Common Errors and Problems   78
5.8 The Future   82

ART TWO ══════════════════════════════════════

RMATS   87
   User Sophistication   88
   'nput Vocabulary   89
     vpes of Formats   90

6.3.1 The X Format   90
6.3.2 X Formats Applied   91
6.3.3 The I Format   92
6.3.4 I Formats Applied   93
6.3.5 The F Format   96
6.3.6 F Formats Applied   97
6.3.7 The A Format   101
6.4 Format Delimiters   103
6.4.1 Commas and Parentheses   104
6.4.2 Commas and Parentheses Applied   106
6.4.3 The Slash   109
6.4.4 The Slash Applied   109
6.5 Review and Practice   117

7   SPSS: VERSIONS 6, 7, 8, AND 9   123
7.1 Basic Structure and Function of SPSS   124
7.1.1 Basic Pathways in SPSS   124
7.1.2 Pathway I   126
7.1.3 Pathway II   129
7.1.4 Pathway III   131
7.1.5 Choosing the Analysis   132
7.1.6 Available Analyses   134
7.2 Input to SPSS   136
7.2.1 The FIXED Format   136
7.2.2 The FREE FIELD Format   137
7.2.3 The DATA LIST Option   140
7.2.4 Tape and Disk Input   144
7.3 SPSS Applied   145
7.3.1 Practice with the FREQUENCIES
          Subprogram   146
7.3.2 Practice with the CROSSTABS and T-TEST
          Subprograms   152
7.3.3 Practice with the SCATTERGRAM and NPAR
          TESTS Subprograms   156
7.3.4 Practice with the REGRESSION and ANOVA
          Subprograms   161
7.3.5 Practice with the DATA LIST Option   165
7.4 Pro Tips and Techniques for SPSS   167
7.4.1 Abbreviations and Plurals   167
7.4.2 Order of Program Control Language (PCL)
          Instructions   167
7.4.3 Syntax of SPSS   169

7.4.4 Data Transformations and Conversions   173
7.4.5 Special Features of Version 9   176
7.5 Pro Tips and Techniques for Specific Subprograms   177
    7.5.1 The CONDESCRIPTIVE and FREQUENCIES
           Subprograms   178
    7.5.2 The CROSSTABS Subprogram   179
    7.5.3 The AGGREGATE Subprogram   180
    7.5.4 The BREAKDOWN, CROSSBREAK, and T-TEST
           Subprograms   181
    7.5.5 The DISCRIMINANT, FACTOR ANALYSIS,
           GUTTMAN SCALE, and CANCORR (Canonical
           Correlation) Subprograms   182
    7.5.6 The REGRESSION Subprogram   182
    7.5.7 The PARTIAL CORR (Partial Correlation)
           Subprogram   183
    7.5.8 The PEARSON CORR (Pearson Correlation),
           NONPAR CORR (Nonparametric Correlation), and
           SCATTERGRAM Subprograms   184
    7.5.9 The NPAR TESTS (Nonparametric Tests)
           Subprogram   184
    7.5.10 The REPORT, RELIABILITY, and SURVIVAL
            Subprograms   184
    7.5.11 The ANOVA and ONEWAY Subprograms   185
    7.5.12 The MANOVA (Multivariate Analysis of Variance)
            Subprogram   186
    7.5.13 The BOX JENKINS Subprogram   187
    7.5.14 The NEW REGRESSION Subprogram   187
    7.5.15 The GRAPHICS Subsystem   188

8   SAS: VERSIONS 79 and 79.5   189
8.1 Basic Organization of SAS   189
    8.1.1 SAS Data Sets   191
    8.1.2 Summary   194
8.2 Specifics of SAS Input   195
    8.2.1 Column Input   198
    8.2.2 List Input   200
    8.2.3 Miscellaneous Input Features   202
8.3 Missing Values and SAS   203
8.4 SAS Procedures   204
    8.4.1 Requesting an SAS Analysis   204
8.5 SAS Applied   206
    8.5.1 Practice with the FREQ (Frequency) and
           UNIVARIATE Procedures   206

8.5.2 Practice with the CORR (Correlation) and TTEST
       Procedures   213
8.5.3 Practice with the CHART and PLOT
       Procedures   218
8.5.4 Practice with the STEPWISE and ANOVA
       Procedures   222
8.6 Pro Tips and Techniques for SAS   226
8.6.1 Abbreviations, Plurals, and Hidden
       Conventions   226
8.6.2 Order of PCL Statements   228
8.6.3 Syntax   229
8.6.4 Data Transformations and Conversions   232
8.6.5 Error Messages   235
8.6.6 Special Formats   236
8.6.7 Special SAS Functions   238
8.7 Pro Tips and Techniques for Specific Procedures   238
8.7.1 The CORR (Correlation), FREQ (Frequency),
       and RANK Procedures   239
8.7.2 The MEANS, SUMMARY, and UNIVARIATE
       Procedures   240
8.7.3 The ANOVA, DUNCAN, FUNCAT (Functions of
       Categorical Responses), and GLM (General Linear
       Models) Procedures   240
8.7.4 The NESTED, NPAR1WAY (Nonparametric One
       Way Test), PROHIBIT, TTEST, and VARCOMP
       (Computation of Variance) Procedures   242
8.7.5 The STEPWISE, NLIN (Nonlinear Regression),
       SYSREG (Systems Regression), RSQUARE
       (Regression with R^2), and GLM Procedures   244
8.7.6 The CANCORR (Canonical Correlation),
       CLUSTER, DISCRIM (Discriminant), FACTOR,
       GLM, GUTTMAN, NEIGHBOR, and SCORE
       Procedures   245
8.7.7 The AUTOREG (Auto Regression), SPECTRA
       (Spectral Analysis), MATRIX, and PLAN
       Procedures   245
8.7.8 The CHART and PLOT Procedures   245
8.7.9 The FASTCLUS (Fast Clustering), PRINCOMP
       (Principal Components), VARCLUS (Variable
       Clustering), and STEPDISC (Stepwise Discriminant
       Analysis) Procedures   246
8.7.10 The RSREG (Response Surface Regression) and
        REG (Regression) Procedures   246

8.7.11 The TRANSPOSE and PRINT Procedures   247
8.7.12 The BMDP Procedure   247
8.7.13 SAS Utilities and OS Utilities   247

9   BMDP: VERSIONS 75, 77, 79, AND 81   249
9.1 Components Common to All BMDP Programs   249
9.2 Input to BMDP   256
        9.2.1 Basic Input Procedure   257
        9.2.2 The FREE Formats   258
        9.2.3 The STREAM Format   260
        9.2.4 The SLASH Format   261
        9.2.5 Summary   261
9.3 BMDP Applied   262
        9.3.1 Practice with Program P4D   262
        9.3.2 Practice with Program P1D   265
        9.3.3 Practice with Program P4F/P1F   273
        9.3.4 Practice with Programs P1R, P1D, and P2V   277
9.4 Pro Tips and Techniques for BMDP   282
        9.4.1 Formats   282
        9.4.2 Abbreviations and Plurals   284
        9.4.3 Order of PCL (BMDP versus BMD) and Program
                Cycling   284
        9.4.4 Data Transformations and Conversions   286
        9.4.5 Data Selection and Checking   288
        9.4.6 Error Messages and Deliberate Errors   291
9.5 Pro Tips and Techniques for Specific Programs   291
        9.5.1 The P4D, P1D, P2D, and P3D Programs   292
        9.5.2 The P7D, P9D, P5D, and P6D Programs   294
        9.5.3 The P4F Program (P1F, P2F, P3F, P8D,
                and PAM)   294
        9.5.4 The P1R, P2R, P9R, PAR, and PLR
                Programs   295
        9.5.5 The P1V and P2V Programs   295
        9.5.6 The P3V, P8V, and P4V Programs   298
        9.5.7 Other Programs   300
9.6 Summary and Comments on BMDP   301

APPENDIXES   303
A. Extended Data Sets for Exercises   303
B. Sources for Statistical Packages and Special
    Programs   308
C. OSIRIS III   308

D. BMD   310
E. SCSS   310
F. BCD TO EBCDIC Conversion   312
G. Data Sets for Student Solutions   313

Troubleshooting Index   317
General Index   320

Preface

Statistical analysis of data is often a very routine, boring task. Fortunately for most social scientists, there are a number of computer systems designed expressly for computer-aided statistical analysis. Despite the fact that such systems can save months or even years of effort, they are complex and densely written, and often seem forbidding to the potential user. The aim of this book is to demystify these statistical analysis systems and to make it easier for a social science researcher to use the computer.

Computer Usage for Social Scientists concentrates on three of the most popular of these systems: SPSS (Statistical Analysis Package for the Social Sciences), SAS (Statistical Analysis System), and BMDP (Bio-Medical Computer Programs). This book is not a statistical analysis system; rather, it is a tool that will facilitate the use of these program packages. Perhaps for many readers it will be the key that unlocks the world of fast, powerful, and labor-saving computer-aided statistical analysis.

This book is organized in two parts. Part One serves as a general introduction to statistical analysis for beginners and as a review for more advanced users. Its chapters on computer language, data collection, and terminals should prove very useful in this context. Part Two focuses on the three analysis packages themselves, featuring general discussion, exercises, and a "Pro Tips and Techniques" section for each. Throughout the text every attempt has been made to distinguish among the needs of beginning, intermediate, and advanced users. In addition, an extensive general index and a special troubleshooting index to the exercises have been provided to facilitate quick reference to specific topics.

Many people helped in the preparation of this book. Special thanks go to Villanova University, especially to the staff of the University Computing and Information Service for their patient and dedicated effort. Professor Daniel J. Ziegler, Chairperson of the Department of Psychology at Villanova, gave advice at all stages, and Cheri Leibold filled the roles of counselor, editor, and unfailing booster.

After an introduction and some definitions of terms, Part One (Chapters 1–5) treats the topics of data collection and preparation, terminals, and keypunches. Strictly speaking, none of these chapters is dependent on the others, but for those students just beginning to learn computer-aided analysis, each chapter will contribute to a solid foundation. Those readers who have used one or more of the major analysis packages for some time should be able to move directly to Part Two.

While it is clear that the future of keypunching and computer cards is limited and that terminals will someday be the standard computer access method, both methods will be used simultaneously for some time to come. It is strongly recommended that anyone just starting out in this field read both Chapters 4 and 5, regardless of the system to be used in actual practice. Although keypunching does not offer the speed and flexibility of terminals, there are valid reasons for becoming familiar with card systems. The pedagogical aspect of cards is most important; they will teach you things about computers that are quite difficult to learn at a terminal.

These first five chapters reflect the author's many years of experience in teaching the basics of computer-aided statistical analysis. As such, they are incomplete and are not the final word on any of the topics discussed.

PART ONE

1

Introduction

The social scientist has long struggled to understand the behavior of individuals and society. One of the social scientist's limitations has always been the massive amounts of data produced by empirical research, not (as many have suggested) the size of the problems themselves. The modern general-purpose computer with its arsenal of programs and procedures is a new element in the struggle. Considerable skill and detailed knowledge are required to manage such a complex and powerful tool.

What are the computer skills required by the social science researcher? The answer to this question is clear. The modern social science researcher need not be skilled in all facets of computer-aided research; usage is the key. The researcher needs only to *use* the computer to further the desired scientific goal. There is little need for expertise in programming, systems analysis, data processing, computer communications, or computer science. The social science researcher needs only to *instruct* the computer as to what is to be analyzed (input), what is to be done with the input (processing), and what is to be done with the results (output). Just as the skilled gardener does not have to be a horticulturist to grow vegetables, the social scientist does not have to be a computer specialist to get results from a computer.

Of course, the psychologist interested in artificial intelligence needs programming skills and perhaps even expertise in fundamental machine logic and number theory. Geographers creating maps of Mars are no doubt more effective if they can work intimately with the computer that makes the whole process possible. However, the sociologist who wants to determine if religious practice is influenced by higher educa-

3

tion need not be interested in random access memory, peripheral storage, processing maximization, do-loops, and similar topics. He or she does need to concentrate on the various aspects of information planning, collection, analysis, and interpretation. When computers can assist, or can increase efficiency along these lines, that is excellent! But the social scientist must guard against allowing the computer to become the central focus. The individual, society, and culture are the issues, while computers are only one of the tools. It is the use of the computer as a tool that is the central theme of this book and the key concept for the social scientist.

Typically, graduate training in computer usage consists of a course or two in the programming language FORTRAN and a large dose of "sink or swim" computer use. The first trip to the computer center often leaves the prospective social scientist with the feeling that computer people are actually from another planet! Methods and statistics courses often merely tease students with vague hints about the computer on the other side of the campus. Academic trauma of this type is not an immutable fact. With today's analysis packages, one can become a skilled computer user.

The skilled researcher of 1960 did not have to be (and most often was not) a user of computers. The central issue at that time was evaluating the relative efficiency of two methods: either one could spend three days running and checking calculations on an electromechanical desk calculator, or one could invest considerable time in becoming a proficient computer programmer and then try to find the time to write and validate a specific program. In the intervening years, we have seen the rise of the general-purpose social science analysis packages written by professionals. These packages have been optimized for speed, cost, and ease of use. It has been estimated that as many as 90% of all social science analysis needs can be met by these packages.[1] Using such a general-purpose program is not much different from executing a computational procedure on an electromechanical calculator or on an up-to-date pocket electronic calculator or even from doing the arithmetic by hand. All of these methods center on a procedure that, when executed correctly and with the right data, will produce the desired result.

It is clear that the contemporary social scientist must learn to use the general analysis packages. The question is how one can become a user of analysis packages. The novice researcher must first become familiar with the structural organization of these packages. A general outline of this organization is as follows:

---

[1] Hugh J. Cline. "Social Science Computing—1967–1972," *Proceedings of the Spring Joint Computer Conference, 1972.* Association for Computing Machinery, 1972, pp. 865–873.

1. Packages are collections of programs.
2. Programs are sets of explicit instructions.
3. Instructions are composed of very detailed steps.
4. Each step is ultimately translated into a binary yes-or-no electrical operation of the computer.

Social scientists typically work at the top of the pyramid and concern themselves only with the global organizations and operations. A typical organization is as follows:

1. Alert the computer that you have work for it to do.
2. Tell the computer which particular action you desire.
3. Tell the computer which options to take for each action.
4. Determine the means of submitting information to the computer (from paper to electrons).
5. Determine how the results should be arranged and the medium to be used (from electrons to paper).
6. Alert the computer when you are finished and no longer require its services.

A farsical example of such computer instructions is shown in the table.

You	Instructions	Computer
Hey you!	WAKEUP	Who's yelling?
I want some stat done.	STAT	Okay, the stat package.
I want averages.	MEANS	Okay, the means sub-package.
I want average IQs for seven fourth grade classes of 27 students each.	GROUP SIZE IS 27	Okay, 27 per group.
The info is on those funny cards with the holes.	CARD	Okay, input is via cards.
Use one card per student.	HOLES 1,2,3	Okay, each data element is on one card in positions 1, 2, and 3 and is ordered by group.
I want each average IQ printed on a separate page.	(PRINT:MEAN IQ, NEXT PAGE) X 7	Okay, output is via printing—one IQ per page.
All right, you can stop.	FINISH	At last.
I'm done.	THE END	Next?

The entire process from the user's standpoint is much like using a cookbook. Follow the instructions carefully, and the result will be a success. Deviate, and the result will be a flop. Unfortunately, using a computer is unlike following the average recipe in that even minor omissions can create major errors. If this text achieves its goals, the occurrence of such errors will be kept to a minimum.

1.1   Computer Availability and Use

Computers and, thereby, computer analysis have become widely available. As the price for a given capacity of information processing has decreased, computers have become affordable to more and more educational institutions. At this point almost all four-year colleges and universities have either an on-campus computer center or special telephone connections to link them to a regional computer facility. The computational power available to the most modest of colleges would have been beyond the reach of even the largest and most pioneering institution fifteen years ago. A researcher working in a home workshop can gain access to large computation facilities for the price of a phone call and two pieces of equipment costing less than half the price of a small automobile.

Not only are computers and computation facilities widely available to the educational community, but they are all fairly general in purpose and function. The single-use computer has all but disappeared except in industrial/research applications and in the military. The capabilities of general-purpose computers are limited only by the skill and imagination of the users. The general-purpose computer of years past was indeed versatile, but it obtained that versatility by executing instructions written in some special language that took months to learn. Today the general-purpose analysis package is so widely available and so easily used that all students, researchers, and educators are potential computer users.

While computers have proliferated, the number of analysis packages has decreased. The reason for this is readily understood; the better organized, highly versatile, and more easily used packages have spread, whereas the other packages have never gained more than local support. The result has been the reduction of the number of packages in nationwide use to a half-dozen or less. The most widely used packages are Bio-Medical Computer Programs (BMD and BMDP), Statistical Analysis Package for the Social Sciences (SPSS), and Statistical Analysis System (SAS). There are perhaps ten additional packages with limited distribution and use (for example, Iowa State University's AARDVARK system, for the analysis of variance problems, or ALICE, for the manipulation and analysis of multidimensional data). There are hundreds of specialty packages (for example, the General Inquirer) and thousands of individual programs. Not one of these requires the user to be a computer programmer.

1.2   Time/Benefit Ratios

One aspect of general-purpose analysis packages often given inadequate consideration is the time/benefit ratio. This is the total of the time required to prepare data for computer analysis and the time required to properly instruct the computer compared to the total time needed to execute the analysis by hand. The ratio does not always favor the computer. For example, to correlate high-school grade averages and college grade averages for 100 subjects would require at most a half hour, including set-up time on a calculator. The trip to the computer center generally requires this much time in addition to the time spent preparing and submitting data. And, of course, one must wait for the results or else make yet another trip. The choice of the calculator is far wiser in this instance.

However, the calculation of even a small matrix of intercorrelations, say 10 measurements on 100 subjects, or cases, is quite time-consuming using a calculator, unless one has some rather special training and is willing to devote years to pushing buttons. Even the data analysis of modest research surveys in political science is laborious to do on a calculator. Many maps would be impossible to construct without computer aid. Clearly then, there is a point beyond which computer analysis is worthwhile in terms of time invested. Unfortunately, such mini/max functions cannot be stated in a general way. Experience is the best guide.

The choice is further complicated by the ever-increasing capabilities of electronic calculators and the spread of telephone connections to large, centralized computers. For the near future, the careful researcher should continue to devote attention to the trade-off, or balance, between quick, immediate hand calculation and slower (but permanent), easily reanalyzable computer calculation.

1.3   Analysis Package Manuals

This book does not replace a relevant analysis package manual, nor does it simply restate such a manual in different language. Rather, it is the overall process that the manuals omit that is of concern here. The package manuals attempt to introduce the novice to the mysteries of computer-aided analysis by a "try it, you'll like it" approach. This is simply a variation of the "sink or swim" approach. Instead of being urged to attempt an analysis without prior instruction (which is like jumping into the deep end of the pool), the reader is urged by the manuals to follow on faith a set of very basic instructions. This is equivalent to wading in a child's backyard pool as an introduction to swimming. Trying it will not necessarily lead one to like it unless there is some preparation. Furthermore, if one tries and fails, negative attitudes are almost certain. Thus, researchers who have not been quantitatively and numerically oriented throughout their academic careers know that there is an extensive and valuable computer world just beyond their reach. The analysis packages alone are useless for a great many researchers. Alone, this introductory text has no purpose. This book and the relevant manual *together* provide a computer-aided analysis system. This book will guide the student or researcher through the seemingly unfathomable language and impenetrable fog of qualifications, exceptions, and outright oversights in the average manual. There are also occasional errors and logical cul-de-sacs awaiting the unwary reader of many manuals. Many of these pitfalls are only avoided after much experience. The experience of the author in this area should prove to be an aid to readers in those critical first steps on the road to the skilled use of analysis packages and their manuals.

1.4   How to Use This Book

The organizational plan of the chapters that follow is one of sequential progression for the novice. Part One (Chapters 1–5) is devoted to topics that many users already comprehend. Part Two (Chapters 6–9) is technical and quite detailed. For this reason, the use of this book requires different strategies for different individuals. In a rough way, users of this book might be categorized as the novices, the intermediates, and the experts. The following suggestions should not be considered mandatory by researchers or instructors. Selected sections or even whole chapters can be skipped without affecting the comprehension of any other section, provided that the skipped material is known from previous study or is covered in lecture. There is nothing sacred about the particular order of the chapters; they may be read as one's situation requires. Those who have not had a course in statistics or analysis methods should not regard this text as a substitute, however.

Novices should progress through the chapters in the order in which they are presented. This will prevent their having to backtrack when errors are made because of insufficient knowledge. Usually novices will be learning to use analysis packages with the aid of a course and/or an instructor. The exercises and examples utilize special short data sets. Expanded versions of these sets are available in a special appendix at the end of this book, and beginning students can use either of these versions.

A student may use his or her own data, but should either seek advice on data preparation or read the chapter on data collection and preparation first. Some collection forms and methods of arranging data for computer use produce a worthless mass of numbers. Redoing several hundred or several thousand data forms and computer cards can be the undoing of a budding package user. All-electronic terminals do not alter this situation; in fact, they accelerate the process. If you intend to use your own data, then you should understand the principles in Part One of this book first.

It may seem unnecessarily dull to spend time and effort on the preparation of data that are fictional and have no meaning beyond that of instruction. Initially, however, this approach is recommended, since researchers too often become enraptured with their own data. ("What does it mean?" "What will the outcome be?") For novices caught up in the rush to determine results, serious errors are the rule rather than the exception. This is definitely one of those situations where once is more than enough. Getting totally useless results is a painful experience. The dullness of artificial data is a virtue in that it directs attention to the task rather than to the excitement of results.

Beyond the pain of redoing an entire data preparation, the novice lives under the constant threat of not even recognizing errors in data. Undetected errors will lead the researcher to erroneous conclusions, and when such conclusions appear in print as valid information entire disciplines may suffer. The researcher cannot afford the risk of disseminating false information. In short, be sure of your ability to prepare data; *then* analyze it.

The intermediately skilled researcher has a different problem: that of increasing skill without unnecessarily repeating previously learned material. The wise course of action would seem to be to determine the level of one's acquired skill and how current it is. If the reader has been away from computer analysis for a year or two, then a brief but thorough review of the opening chapters of this book will be helpful. Even if one is fairly confident of one's skill, reestablishing the vocabulary and perspective required in computer-aided analysis will be beneficial, particularly when problems are encountered. Remember–the professional computer consultant, though highly skilled, speaks a specific language. It is a language that is different from, and not easily translated into, the language of the social scientist. Unless one possesses an unusually high tolerance for frustration, some review is advised.

The intermediately skilled user whose knowledge is current should study only those chapters that are devoted to topics outside of present skills. The use of actual research data instead of the artificial data from the special appendix is quite appropriate in this case. For example, if a researcher has been successful in using a specific analysis package and wishes to learn to use another package, then it is most logical to utilize data previously prepared. Even within a specific analysis package, there is much to be learned about modifying, reorganizing, and storing data that can be expedited by using "old" data.

Finally, knowledge of particular subsections of an analysis package does not automatically imply skill with all sections. The intermediately skilled researcher would do well to use the text as an aid to discovering those aspects of a favorite package that have not as yet been used. Also, intermediately skilled users with specific gaps in basic knowledge (such as not really knowing how to run a terminal or a keypunch) may find this text a means of filling those gaps. Many researchers simply hire someone to prepare their data for them and thus never truly comprehend the requirements of the task. It is still painfully apparent that the only way to correct the errors of others is to understand, at least in a general way, the specifics of the task. This is particularly true in data preparation, and intermediate users can consider this text as an opportunity to remedy the situation.

Expert users are likely to require this text only in the contexts of easy entry to new analysis packages and of specific review. It is easy to forget many details; for example, advanced users are well aware that the outdated IBM 026 keypunches are still quite serviceable for data but not for special symbols unless certain corrections are made. What are those corrections? For the advanced user this text can provide a ready source of that information (see Chapter 5 and Appendix F).

Experienced users of the statistical analysis packages will probably find the "Pro Tips and Techniques" section of each of the analysis package chapters of this book to be of greatest interest. In these sections, the author has summarized 16 years of experience with these packages.

For users at all levels, the computer terminal is the wave of the future. Data entry via cards will become obsolete, and direct data entry from a terminal will be the standard. Many larger institutions are already undergoing this changeover. Keypunches still remain in use, especially in smaller institutions. At some point, however, terminals will be the norm.

At present, standardization is the rule for cards, but terminals have not yet reached this point. The terminals come in an amazing variety, and the ways in which they are connected to the computer also vary widely. Chapter 4 is devoted exclusively to terminals and should be reviewed by those readers who are not familiar with their use. Throughout this text, a line of data will be treated in the same manner whether input is via cards or terminals.

2

Computerspeak

A dictionary of sorts appears at this point to familiarize you with some of the basic terms used in the world of computers. Its length has been kept to a minimum; only terms essential to social scientists have been included. Hundreds of additional terms exist but have been omitted, since the primary goal of this book is to illustrate and to involve you in the principles of statistical analysis. The definitions are intentionally short. The typical social scientist does not need to know that the hexadecimal code is a numeration system with a radix of 16 or why this is advantageous. Simple recognition in a computer printout of hexadecimally coded information is sufficient. No doubt many social scientists know and use a large computer vocabulary, but it is hardly a requirement.

The chapter is divided into five parts: hardware, software, codes, acronyms, and old words–new meanings. In order to communicate with the professionals at the computer center, it is necessary to know the basic terminology used in their world. But it is not at all necessary to memorize these definitions; you will know them like the back of your hand quite soon. By the end of this book, you will realize that the biggest obstacle is not the meanings of the terms themselves but the maze of interrelationships among them.

2.1 Hardware

hardware: The computer and all its accessories—in other words, the machines. The associated term *hardwired* refers to the many

13

special-purpose devices that can be created, such as machines that add, read documents, answer phones, etc. Theoretically, a machine could be constructed to perform any definable task. Most computers are general-purpose machines, although many of their component parts are not.

central processing unit (CPU): The computer itself. All calculations and decisions occur here. Bulk memory, input, output, and communication are handled elsewhere. CPUs come in all sizes and differ widely in operation and design. The attached memory is often called primary storage. The average social scientist is oblivious (appropriately) to the inner details.

primary storage: The all-electronic memory that is part of the central processor. Even a simple instruction such as "ADD 4 TO 6" cannot be executed unless it is in primary storage. Everything that is on disk, tape, etc., must first be transmitted to primary storage before any instruction can be executed.

microprocessor: A CPU in a miniature form without memory. For the social scientist, the only significant difference is size and capacity.

card (computer card): A paper card of varying dimensions with punched holes arranged in a specific pattern. The pattern for the holes depends on intended use, the most common being BCD (Binary Coded Decimal). BCD is nearly universal in social science use. The point is that a computer card contains information or instructions that can be read by a computer (computer-compatible).

keypunch: A device for punching holes in computer cards. It utilizes a typewriterlike keyboard. Generally, keypunches also print the meaning of each hole along the top edge of the card. (See Chapter 5.)

card reader: A device attached to the computer that reads the information from the cards into the computer. In some installations, the user simply hands the cards to the computer-center staff; in others, the user actually places the cards into the reader.

printer: A device that prints the results of a computation or computer activity onto paper. (The paper is usually called output.) This is where the computer communicates with the users. Printers vary from slow typewriterlike devices to high-speed, high-technology monsters. They are not necessarily limited to printing only in uppercase characters. So far it has simply been cost-efficient to do so, although this is changing rapidly.

terminal: Another device allowing the computer to communicate with the user by incorporating a televisionlike screen or a typewriterlike printer with a keyboard. A terminal is a two-way communication device, since it also serves as a means of input. Terminals are replacing cards as the most common input procedure.

modem: An acronym derived from modulator-demodulator. Modems generate tones that form coded signals that allow the terminal to communicate with a computer. Acoustic modems (also called acoustic couplers) use a phone handset, while high-speed modems are directly connected to the phone system. (See Chapter 4.)

disk (also spelled disc): A platter of magnetic material on which information is stored in bulk. This form of memory is slower than primary storage but is faster than tape.

tape: A thin ribbon of magnetically charged plastic, wider than audio tape, and coded digitally for computer use. It is stored on large reels and provides slow but inexpensive memory.

punch: An output device producing punched cards instead of printed pages. Information is coded, and unlike a keypunch, the cards generally do not have printing along the top. Computer-generated cards may be processed by an interpreter or special keypunch to decode (print) the holes.

paper tape: Tape with holes punched in it that serves the same purpose as magnetic tape. It is slower and is more commonly used in laboratory and home computers.

floppy disk, diskette: A smaller, more convenient disk that operates basically like a standard disk. It is usually used for fast bulk storage for microprocessors (small computers).

cassette, cartridge: Reel of tape in a container, usually used with microprocessors.

format: In regard to hardware, the information structure of a device. Thus, there are tape and disk formats, card formats, printer formats, etc. Formats vary not only among devices, but also among manufacturers. Data that are structured for one computer center or manufacturer are likely to be incompatible with those for another. (This is not an insolvable problem, however.)

daisy wheel: The printing element in certain typewriterlike terminals.

dot matrix: A printing element like a daisy wheel but forming characters with patterns of dots.

bit: The smallest unit of operation for digital computers. A bit, or binary digit, typically assumes a two-valued form, i.e., on/off, 0/1, or stop/start. In contrast, a decimal digit would have ten possible states. Generally, computers are assembled from millions of electronic parts that are capable only of this basic on/off operation.

byte: A combination of from 4 to 32 bits that is treated as a unit. Thus an 8-bit byte is a string of eight adjacent binary digits. Depending on size, a given byte could be a letter, a decimal digit, etc. A 1M-byte memory is thus able to store one million bytes of a fixed size.

word: A combination of bytes (also bits) treated as a unit. It is a symbolic representation of an instruction, number, name, etc., that is a basic unit of operation for a specific machine. Just what a word is depends upon the design and purpose of the machine in question.

time designators:

a millisecond = one thousandth of a second.
a microsecond = one millionth of a second.
a nanosecond = one billionth of a second.
a picosecond = one trillionth of a second.

capacity designators:

K, 1K, kilo = one thousand. (One might hear the expression, a 256-kilobyte memory.)
M, 1M, mega = one million.

2.2   Software

The vast majority of computer users do not direct the fundamental operations of the machine itself, the electronic impulses in shifting patterns of 0's and 1's. Some higher-level method of communication (programming language) is used instead. This language functions as a translator that determines the machine operations necessary. Often as not, there is a series of translations. These higher-level languages constitute software. A set of instructions to calculate the gross national product is a software program. Software ranges from the very general and multipurpose to the extremely detailed and single-purpose. SAS, SPSS, and BMDP are at the top of this translation chain and are known as statistical software. The following definitions are the bare bones of this large and complex topic. As before, the explanations are short and include only those details essential to social scientists.

programming languages: High-level instructions, such as "X = A + B," that translate into the machine language needed to execute a given operation. Some of the more popular languages are as follows:

FORTRAN is the old standby for general as well as scientific uses.

COBOL is the standard business language.

PL/1 stands for Programming Language 1, a general-purpose language.

ALGOL is a mathematical language.

APL stands for A Programming Language, a symbolic, general-purpose language.

BASIC is a general-purpose language commonly used with microprocessors.

PASCAL is a general-purpose, up-and-coming language.

Thus, to program the computer is to write a series of instructions in one of these languages, and the instructions themselves are a program.

compilers, interpreters, and assemblers: Programs that translate the high-level programming language instructions into the machine language of 0's and 1's. While a data-processing expert would be appalled at the implication that compilers, interpreters, and assemblers are the same, the social scientist only needs to know their functional definitions in a general way. There are FORTRAN compilers, ALGOL interpreters, etc. Thus, the user's question "Do we have FORTRAN available?" is answered by "Yes, we have a FORTRAN(IV)G compiler."

machine language: The most elementary of programming languages. It does not have to be translated, since it can be said that the computer speaks this language. Only superspecialists use it.

operating system, monitor, executive: The system, by whatever name, that allocates resources, defines who can use the system and when it may be used, and controls other housekeeping functions, such as who is spending what. No general rules can be stated, since each computer-center staff generally modifies the system to meet local needs.

input: Any of the many ways in which information (data) can be placed in the computer. The beginning user typically starts with the punched card or the terminal (other forms include tapes and disks).

output: Any of the ways in which information (data) can be transmitted from the computer to the user. The printed page and the lines of "electronic" print on a terminal screen are typical.

2.3   Codes

Although a special section on computer codes would usually appear only in an advanced text, experience indicates that having a general conception of what codes are about is useful. There are occasions when computer professionals can be unintentionally confusing while talking to nonprofessionals.

binary: The basic two-state logic (0's and 1's) used by modern digital computers. Beyond this broad category, nothing is standard, for there are a huge number of binary codes and formats. A common misconception is that binary assures compatibility. It does not.

BCD: The Binary Coded Decimal. This is a widely used means of representing decimal data (and alphanumeric data) with binary numerals. The actual pattern of 0's and 1's, or code, has many variations. The standard computer card is one such coding scheme. (See chapter on keypunches.)

EBCDIC: The Extended Binary Coded Decimal Interchange Code. This is a newer variation of BCD, used by IBM.

ASCII: The American Standard Code for Information Interchange. This is the standard communications code for non-IBM equipment. (See chapter on terminals.)

Baudot (baud): The standard teletype code, of use only when using a teletype with other equipment. A *baud* is a measure of communication speed.

card: A more or less standard code for punched cards. The form of 12 rows by 80 columns is typical, but there are others. The meaning of the pattern of the holes obviously can vary as well.

octal: A pattern of binary encoding based on a string of eight numerals. Social scientists only need to know that octal output from a computer must be decoded.

hexadecimal: Another coding pattern similar to octal except that it is based on units of 16. The letters A through F represent the decimal numbers 10 through 15.

2.4   Acronyms

For reasons known only to the gods, computer professionals love acronyms. Here is a brief list of some of the more common acronyms:

CDC: Control Data Corporation, one of the "other" computer manufacturers.

CPU: Central Processing Unit, the computer itself.

CRT: Cathode Ray Tube, or televisionlike screen. The acronym is a leftover from the early days of vacuum-tube technology.

DEC: Digital Equipment Corporation, a major manufacturer of computers.

PDP: Digital Equipment Corporation's name for their popular line of computers. There are PDP-8s, PDP-11s, etc.

JCL: Job Control Language, an IBM term for the instructions given to the operating system.

SCL: System Control Language, an acronym invented by the author of this book to be used as a substitute for the IBM term JCL. It refers to *any* operating instructions that control what the computer does in terms of selected programs, tapes, and packages.

PCL: Program Control Language, another invented acronym that refers to those instructions that control a specific statistical package.

OS: Operating System.

2.5   Old Words—New Meanings

A number of common words have become such an integral part of the vocabulary of the world of computers that the user of a statistical package should be acquainted with their meanings in this context.

merge: The combining of two or more files, or other organized information, into one.

submit: The presentation of a piece of work to a computer or computer center.

sorter: An electromechanical device for sorting computer cards (these are fast fading into oblivion).

record: A unit of information. For a standard card, a record is 80 characters (maximum). The size of a terminal record depends on the width of the screen. With tape and disk, records are of variable lengths.

deck: A group of computer cards. A data deck is a pack of cards on which a user has recorded research data.

port (synonymous with data channel): The pathway and/or devices allowing input and output operations.

file: An organized set of information retained on cards, a tape, a disk, or through a terminal system.

data base: A large set of files, such as a warehouse inventory or a group of surveys.

data bank: A very large body of information, such as the census or IRS information.

run (or computer run): The process of (1) submission of data and instructions to the computer, (2) the actual execution of the in-

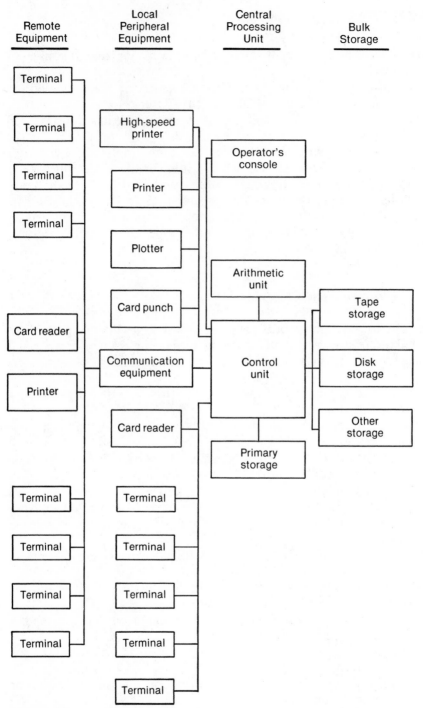

FIGURE 2.1. The Structure of a Modern General-Purpose Computer

20

structions by the computer, and (3) retrieval of the results on paper, cards, or terminal lines. In IBM installations one would hear, for example, "My first run of the *t*-test program was successful."

2.6   Structure

The social scientist can remain blind to many aspects of the structure of the computer, but a limited degree of familiarity makes communication with computer consultants and other professional personnel far easier.

Figure 2.1 is a schematic of a modern computer system. It contains the essence of what the social scientist needs to know. Obviously, some computers are more complex than this, and many are far simpler. While there is no need for a detailed discussion of the diagram, a few clarifications are in order. The heart of the system is the CPU. Everything else is subsidiary to it. More important than the various components is the communication system. In the CPU all communication is two-way, but many other connections operate in only one direction. One-way connections are either from input or to output devices—not both. Thus, a card reader is only an input device, whereas a tape drive is both an input and an output (two-way) device. Printers are one-way output devices. This is an important concept because the all-time classic beginner's mistake is to attempt to input on an output device or vice versa. The lesson is simple enough: read on readers and print on printers.

Equipment that operates in the local computer environment can also be operated remotely. Most often this is done over the public phone system. This means that a student/researcher in one part of a campus can work with a computer in another area. (Users in Hawaii can work with computers in Washington, D.C., via satellite.) The only limits to this are the price of the phone call and the cost of computer time. Remote systems also operate at considerably slower speeds. High-speed telecommunications are possible but require leasing special data transmission phone lines. Generally, local operations provide more options, greater speed, and lower costs.

3

Data Collection and Preparation

Although we have all heard of someone who successfully records the results of scientific research on the backs of old envelopes, most of us do better with a more structured approach. This is particularly true in computer-aided statistical analysis, for the machine that can read scrawls on envelopes has yet to be invented. In addition to the machine-oriented, structural demands, the ease, accuracy, and speed of data collection and preparation should be maximized.

The following guidelines for data collection and preparation summarize the major problem areas:

1. The complete process of collection, preparation, and entry should be planned prior to doing any actual work. While this would appear obvious, the urge to get to the heart of the research is often strong and can produce data in a form that cannot be analyzed by any statistical package. Then it could take weeks to rework the data into a suitable form. The alternative, doing the statistics with a calculator, is just not viable in many cases. Researchers who have imposed this method on themselves have often regretted it.

2. Collection forms and procedures should be designed to keep the number of steps to a minimum. One should avoid collecting data in some convenient manner, then transcribing data onto a special coding form, and then transcribing them again onto the actual computer input medium. (The medium traditionally has been

cards; however, terminal entry is becoming quite common.) Over-all, this three-stage process is a waste of time. A far better proce-dure is to design the original data collection forms and procedures so that the middle step is eliminated. The streamlining of such procedures is the heart of this chapter. There is no single style that allows for all research situations, so several examples are given. All can be adapted to specific situations.

3. All data entry systems are sequential and systematic. Data should be collected with this in mind. Packages described in this text and the majority of all other packages use the standard subject-row data arrangement (see Section 3.1). This may seem obvious, and yet many of the common problems brought to this author could have been avoided with a little reading.

4. Keep the scoring procedures consistent. Do not change from

    Yes _____ , No _____ , N/K _____

    to

    No _____ , Yes _____ , N/K _____

    This author recently came across one study with at least 17 differ-ent style changes in it. This may serve to keep respondents awake, but will surely produce preparation problems. (There *is* a type of test in which style changes are deliberate. If you are using this variety, then machine-readable, or mark-sense, forms are the only solution.)

5   It is essential to read the description of the statistical program to be used before planning the collection procedures. Many statis-tical programs function correctly only when the data are entered in specific ways. Both card and terminal entry have fixed require-ments. Cards hold a maximum of 80 characters, and the informa-tion must be arranged systematically. Terminals also require systematic arrangement of information, and as yet there is no standard size. (An introduction to cards, terminals, and the stan-dard data setup is given in Section 3.1.)

6. When the data will be prepared for the computer by professionals, it is best to negotiate the details before the data are collected. Many data-processing facilities will prepare data only if they are in a specific form. Discuss the requirements beforehand.

7. Recording forms of all varieties can be misplaced, shuffled, or separated from each other. A code or identifier that appears on every page of a survey ensures against this. It is very common to reanalyze old data in the light of new evidence. It is painful to know that the answer is somewhere in the data but that the original forms are so scattered and mixed as to make reanalysis impossible.

8. The needs of the research should take priority over the needs of the computer analysis system. This may, in extreme cases, preclude computer analysis. This is the classic mini/max dilemma: how to maximize the research effort and minimize the secondary efforts. On the other hand, some studies are beautifully organized and eminently analyzable but are relatively meaningless in terms of knowledge gained.

All of these principles can be summarized in the aphorism "Look before you leap." Courses in research and research methodology encourage the student to get out there and collect some data. This then brings the students to the consultant with the complaint that they are unable to get SAS, SPSS, or BMDP to analyze their data. Often the problem can be directly traced to insufficient thought about data collection and preparation.

3.1 Standard Data Format—The Matrix

A discussion of data would not be necessary if the nature of computers and computer input (entry) devices placed no constraints on the arrangement of the data. This is not the case. First among these constraints is that many computers, through the medium of a computer program, are most efficient if the information is arranged in a matrix, i.e., a row-by-column (lexigraphical) table. The conventional arrangement of such a matrix is subjects by variables. Consider a study in which the subjects (cases) are rats and the variable under analysis is morphine consumption per day (dependent variable). The data should be arranged in the following way: X represents the actual data, and the associated numbers (subscripts) indicate the position of that data in the matrix. Table 3.1(a) presents this basic arrangement.

Data from a typical experiment on morphine addiction are arranged in Table 3.1(b) in the standard format. Note that all the data for a single subject are contained within a single row of the matrix.

Even if this experiment were more complex, the same data layout would be used. For example, let us say that instead of just one group of subjects the experiment utilizes two groups of animals: high initial addiction and moderate initial addiction. In standard format, the matrix would then be as shown in Table 3.2(a) and (b). (If investigations contain three or more variables, there are alternative forms available to the user of the standard computer analysis packages.)

In the morphine addiction example, it might be desirable to analyze the information in a pairwise fashion, i.e., to pair subject 1 from

TABLE 3.1. Standard Data Format

Part (a): Standard matrix data layout

		Variables (days)			
		X_1	X_2	X_3	$X_4 \ldots X_p$
	1	X_{11}	X_{12}	X_{13}	$X_{14} \ldots X_{1p}$
	2	X_{21}	X_{22}	X_{23}	$X_{24} \ldots X_{2p}$
	3	X_{31}	X_{32}	X_{33}	$X_{34} \ldots X_{3p}$
Subjects (rats)	4	X_{41}	X_{42}	X_{43}	$X_{44} \ldots X_{4p}$
	.				
	n	X_{n1}	X_{n2}	X_{n3}	$X_{n4} \ldots X_{np}$

Part (b): Representative data in standard matrix form

		Variables (days)			
	1	20.1*	20.3	19.2	18.4	3.1
	2	21.2	19.3	15.9	11.1	2.4
Subjects (animal #)	3	19.8	19.9	16.3	10.1	3.0
	4	19.2	20.1	17.8	10.0	2.9
	5	20.00	18.9	16.9	11.2	3.2

*ml/kg of body weight

TABLE 3.2. Two-Group Data Format

Part (a): Typical two-group data in standard matrix format

G_1 (treatment group)

		Variables (days)			
	1	X_{111}	X_{112}	X_{113}	X_{114}	X_{115}
	2	X_{121}	X_{122}	X_{123}	X_{124}	X_{125}
Subjects (rats)	3	X_{131}	X_{132}	X_{133}	X_{134}	X_{135}
	4	X_{141}	X_{142}	X_{143}	X_{144}	X_{145}
	5	X_{151}	X_{152}	X_{153}	X_{154}	X_{155}

G_2 (treatment group)

		Variables (days)			
	6	X_{211}	X_{212}	X_{213}	X_{214}	X_{215}
	7	X_{221}	X_{222}	X_{223}	X_{224}	X_{225}
Subjects (rats)	8	X_{231}	X_{232}	X_{233}	X_{234}	X_{235}
	9	X_{241}	X_{242}	X_{243}	X_{244}	X_{245}
	10	X_{251}	X_{252}	X_{253}	X_{254}	X_{255}

TABLE 3.2. (continued)

Part (b): Representative data for two groups

G₁ (treatment group)

		Variables (days)			
	1	20.1	20.3	19.2	8.4	3.1
	2	21.2	19.3	15.9	11.1	2.4
Subjects	3	19.8	19.9	16.3	10.1	3.0
	4	19.2	20.1	17.8	10.0	2.9
	5	20.0	18.9	16.9	11.2	3.2

G₂ (treatment group)

		Variables (days)			
	6	21.3	16.1	9.7	3.0	0.9
	7	19.8	17.3	10.4	2.7	0.8
Subjects	8	21.0	19.0	11.2	3.3	1.2
	9	20.2	16.7	8.7	2.3	1.0
	10	19.7	16.3	9.9	3.1	0.7

Part (c): Representative two-group data in alternative standard form

		Variables (days)			
	1	20.1	20.3	19.2	8.4	3.1
	6	21.3	16.1	9.7	3.0	0.9
	2	21.2	19.3	15.9	11.1	2.4
	7	19.8	17.3	10.4	2.7	0.8
Subject pairings	3	19.8	19.9	16.3	10.1	3.0
	8	21.0	19.0	11.2	3.3	1.2
	4	19.2	20.1	17.8	10.0	2.9
	9	20.2	16.7	8.7	2.3	1.0
	5	20.0	18.9	16.9	11.2	3.2
	10	19.7	16.3	9.9	3.1	0.7

group G₁ with subject 6 from group G₂, etc., resulting in five pairs. Table 3.2(c) is arranged in this way. Analysis packages vary among themselves as to which of these arrangements is necessary. In the strict sense, both are standard formats; that is, all data for an individual subject (animal, voter, parent, etc.) are grouped in a row, not a column.

TABLE 3.3 Data from Table 3.1(a) Arranged in a Subject-Column Matrix

		Subjects (rats)				
	1	20.1	21.2	19.8	19.2	20.0
	2	20.3	19.3	19.9	20.1	18.9
Variables (days)	3	19.2	15.9	16.3	17.8	16.9
	4	8.4	11.1	10.1	10.0	11.2
	5	3.1	2.4	3.0	2.9	3.2

When a column arrangement is used, as in Table 3.3, the investigator is confronted by a cumbersome problem. No widely used analysis package will directly accept data in subject-column form. It is very important to realize that a computer analysis package cannot, under most circumstances, differentiate between data presented in subject-row and those in subject-column form. If we assume that the original data in Table 3.1(b) show a significant tendency for morphine addiction to decrease after the subjects received a morphine antagonist prior to the first day, it is easy to observe in Table 3.3 that there is no similar decrease in morphine consumption. Actually, quite the opposite conclusion would be reached: that individual subjects differ in the amount of morphine consumed but that they do not decrease consumption over time. Thus, an inadvertent subject-column ordering of the data would be a very serious mistake.

To prevent such incorrect analyses, would it not be possible to indicate to the analysis package the particular form of each set of data (subject-row or subject-column)? Yes, it could be done, but such a process would require a complete analysis package written with parallel but opposite orientation. The other alternative would be to write a special program exclusively for the purpose of changing subject-column formats to subject-row formats. The first alternative is extremely wasteful of computer storage, and the time requirements to create such a package are huge. The second alternative does exist; there are matrix transposition routines. However, they are very slow and laborious and require huge amounts of computer memory (and therefore money). The only reasonable approach is to plan to meet the subject-row requirement before data collection begins.

3.2   Data Collection

Although most texts present only good data collection procedures, this one will first present an example of a poorly planned survey research form. Forms that are not well planned are very common, and Figure 3.1 is a case in point. Here is a list of the errors illustrated in this example:

1. The responses are scattered, thus the correct sequence is not obvious.
2. The scattered responses will interfere with the actual physical preparation of the data because the preparer's attention must move back and forth, rather than down a single column.
3. The ultimate physical arrangement of the data is not obvious from the form.
4. The form is not sufficiently identified. For example, whose study is this? What is a CD?
5. The range and the possible variation of scores are not clear.
6. If the subject rather than the researcher is to fill out the form, the allowable responses should be stated. With fill-in forms, one must be prepared to receive answers such as "perhaps," "maybe," "50/50," etc.
7. When response order varies, respondents, researchers, and data preparers are all likely to become confused.
8. When the data must be in fixed position with respect to a terminal line or computer card, it is best to place the columns on the form.

In contrast, Figures 3.2 and 3.3 illustrate good data collection forms. In Figure 3.2, the instructions for conducting the experiment are in the left-hand column as an aid to the actual administration of the study. Instead of collecting the subject responses on a separate form and then transferring them, the researcher has the subject respond directly on the final form. When it is undesirable for the subject to see the instructions, a hospital-style covered clipboard, with the lower right portion removed, can be used. Figure 3.2 is not definitive but merely suggests good procedures.

Figure 3.3 is typical of a good survey form. The respondent can circle his or her choices, or the interviewer can circle the choices. With a simple survey and after lots of practice, the interviewer could also code the response into the column at the far right while the interview is being conducted. Be cautious about doing this, for concurrent recording and coding place a great responsibility on the interviewer, since they can lead to unreliable data. The columns of 0's and 9's are typical codes for no answer (refusal to answer) and "don't know." It is unwise to expect the respondent to use the 0 or 9 convention.

Figure 3.3 clearly requires two records per respondent to maintain the implied spacing. Note that each record has three identifiers: an

CD SURVEY

ID _____

Interviewer _____ Date _____

Age _____ Sex _____ Rural _____ Urban _____

State _____

1. Would you vote as a conservative or liberal?

2. Who _____ is your choice for president?

3. Are you registered as a Democrat, a Republican, or other?

4. Did you vote in the last Presidential election? The most recent state election? The most recent county election? The most recent municipal election?

_____

_____

_____

_____

_____

_____

_____

_____

_____

_____

_____

_____

_____

_____

5. Would you categorize the other members of your family as conservative, liberal, or other?

    1. _____

    2. _____

    3. _____

    4. _____

    5. _____

FIGURE 3.1. A Poorly Considered Survey Research Form
(for Completion by the Interviewer)

MEMORY EXPERIMENT 7A

Subject name _____

Student number _____

Test date _____

Participation credits—class _____ Time _____

Instructor _____

Experimenter ID _____

A. Collect above information.
B. Assign/collect 1–4.
C. Give instructions;
   guide practice.
D. Give list of words; start timer.
E. Halt after 2 minutes; collect.
F. Give math sheet; start timer.
G. Halt after 3 minutes; collect.
H. Give response sheet;
   start timer.
I. Halt after 2 minutes; collect.
J. Debrief and thank subject.

Response code: 0 = no answer
                1 = incorrect
                2 = correct

1. Subject ID _____ (1–3)
2. Group number _____ (5)
3. Gender _____ (7)
4. Age _____ (9–10)

Answers

1. _____:____ (13)
2. _____:____ (14)
3. _____:____ (15)
4. _____:____ (16)
5. _____:____ (17)
6. _____:____ (18)
7. _____:____ (19)
8. _____:____ (20)
9. _____:____ (21)
10. _____:____ (22)
11. _____:____ (23)
12. _____:____ (24)

FIGURE 3.2. *Example of a Well-Considered Experimental
            Research Form*

ID number (columns 1–4), a record number (column 77), and the researcher's initials (columns 78–80). This is insurance against inadvertent
scrambling of files if someone drops the cards or sheaves of code forms.
(There is an automatic technique for inserting initials in the last three

CANDIDATE PREFERENCE SURVEY
MARCH 1980

Please rate the candidates using the following scale.

1 = very little	ID _____ (1–4)
2 = little	Geo. area code _____ (6–10)
3 = moderate	Interviewer _____ (11)
4 = a good deal	Female (1) Male (2) ____ (12)
5 = a great deal	Date __,__/__,__/__,__ (14–19)
	\ \ \
	yr. mo. day

QUESTIONS

1. Leadership abilities
   A. candidate 1   1 2 3 4 5 0 9   ____ (21)
   B. candidate 2   1 2 3 4 5 0 9   ____ (22)
   C. candidate 3   1 2 3 4 5 0 9   ____ (23)
   D. candidate 4   1 2 3 4 5 0 9   ____ (24)
   E. candidate 5   1 2 3 4 5 0 9   ____ (25)
2. Foreign policy administration
   A. candidate 1   1 2 3 4 5 0 9   ____ (31)
   B. candidate 2   1 2 3 4 5 0 9   ____ (32)
   (. . . and so on, through)
6. Ability to improve quality of life
   A. candidate 1   1 2 3 4 5 0 9   ____ (71)
   B. candidate 2   1 2 3 4 5 0 9   ____ (72)
   C. candidate 3   1 2 3 4 5 0 9   ____ (73)
   D. candidate 4   1 2 3 4 5 0 9   ____ (74)
   E. candidate 5   1 2 3 4 5 0 9   ____ (75)
7. If you could vote for all candidates, in what order would you prefer
   them?
   A. candidate 1   1 2 3 4 5 0 9   ____ (21)
   B. candidate 2   1 2 3 4 5 0 9   ____ (22)
   C. candidate 3   1 2 3 4 5 0 9   ____ (23)
   D. candidate 4   1 2 3 4 5 0 9   ____ (24)
   E. candidate 5   1 2 3 4 5 0 9   ____ (25)

Card number ____ (77)
DMK ____ (78–80)

*FIGURE 3.3. A Well-Considered Survey Research Form
(for Completion by the Interviewer)*

columns. See Chapter 5.) The record number is essential because many statistical packages assume cards or terminal lines are in the correct order. It is the researcher's responsibility to ensure this order. In a broad sense, all additional data should maintain the pattern established by the first line (card). When the locations change with each card, the processes of recording, coding, preparing, and proofreading become much more difficult.

Figure 3.4 presents a generalized data form. It may be used as an original collection form by labeling the appropriate columns and blanking out any unused ones. The forms will fit into a looseleaf binder. This author has copyrighted the form, but you may prepare a similar version for your own use. Note that the (A) side is meant to be the left-hand page, and the (B) side is meant to be the right-hand page in an open looseleaf notebook. It may be used as an intermediate form. At first glance, it may seem awkward compared to the typical FORTRAN coding form; however, the FORTRAN form has a 72-column limit, asks for a good deal of irrelevant information, and is awkward to use because it is oversized. The computer researchers on a campus can always be recognized by the sheaf of dog-eared forms projecting from their notebooks. The sample form given here measures $8\frac{1}{2}$ by 11 inches.

Two other techniques deserve mention in this context. The first is the use of mark-sense forms. These are the ubiquitous answer sheets with which most students are familiar. These forms are machine-readable, which saves time in terminal or card preparation. Proofreading requirements are also reduced. However, mark-sense forms are difficult to use as the principal collection form unless the population of interest (students) has already been trained in their use. All things considered, machine-readable forms are attractive when dealing with a large volume of information. However, the details of the machine system tend to distract the researcher.

The future of data collection for many researchers is in the interactive computer system. It does not matter if this is linked with a large computer, or an independent miniprocessor or microprocessor. The point is that it is now possible to program an entire experiment or survey so that the subject/respondent sits at a terminal and types responses to questions or to an experimental situation. In a memory experiment, the words to be memorized appear on the screen for one minute, for example, and the subject is then asked to type the words he or she remembers. In survey research, the questions would appear on the screen, and the respondent would type the answers. Clearly, many of the same planning requirements for conventional data collection also apply in this situation.

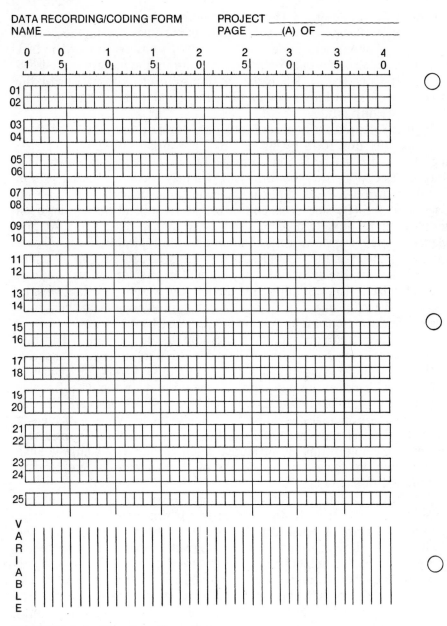

FIGURE 3.4. Two-Part Data Form

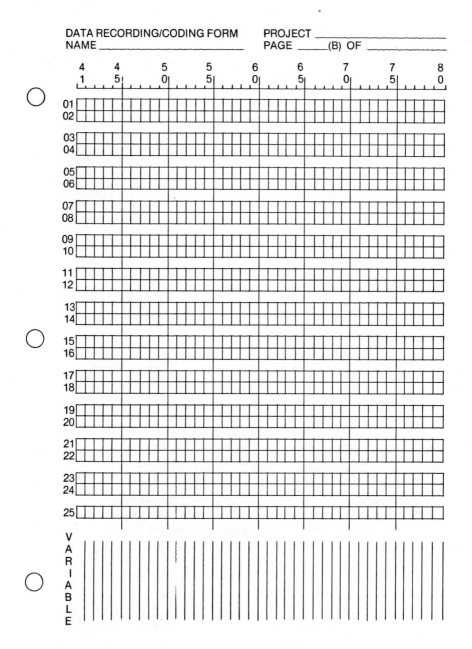

Two additional burdens are associated with this automated data collection. First, the subject will often be naive with respect to terminals, necessitating special instructions. Second, the programming that makes automated data collection possible should include structuring the data in a form compatible with its final analysis by the statistical analysis package. Although it is possible for the minicomputer or microcomputer to calculate the statistics itself, versions of SPSS and other analysis packages for such machines do not exist, and it is a big job to attempt to write them. It is possible with existing technology and will be quite common in the future to have the local minicomputer or microcomputer communicate with the remote large computer via telephone.

3.3   Data Preparation

In order to plan data collection so that it can be easily and efficiently input, the researcher must be familiar with the rudiments of terminals and cards. A more comprehensive treatment of these is given in Chapters 4 and 5.

Figure 3.5 depicts a standard computer card. The small rectangles represent the holes that are physically punched into the card to designate each character. Five questions from a survey could be transferred to a card as shown in Figure 3.6. We assume the respondent answers Y, Y, Y, N, K. This information could have been placed into any group of five columns, every five columns, or every other column as in Figure 3.5. Many patterns are possible; the answers could even be scrambled relative to the order in which they were answered (not a recommended method). The pattern then repeats for every respondent, as in Figure 3.7.

When the answer to the first question always appears in the same position, the data is in a fixed format. Free formats, by contrast, allow multiple subjects per card. The first question might appear in column 1 for subject 1, in column 6 for subject 2, in column 11 for subject 3, and so on. This is still standard subject-row format. For the present discussion, fixed format will be used, and looking at the details of free formats will be delayed until Chapter 6.

Data preparation for terminals is essentially the same, with the screen replacing the cards. It is important to remember two things about terminals: row width is not yet standardized, and each screen can hold many lines. Each line, of course, corresponds to a card. Figure 3.8 is identical to Figure 3.7 except that terminal lines are substituted for cards. Note that when terminal lines represent respondents, the top

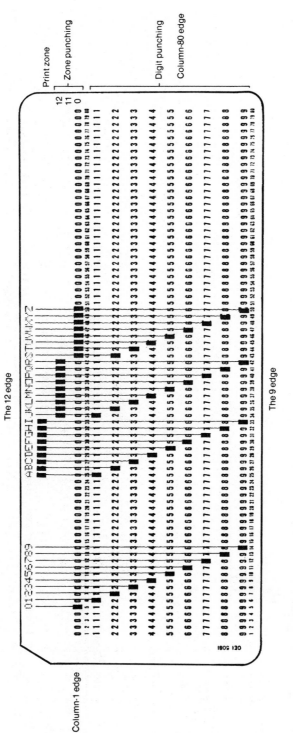

FIGURE 3.5. *The Keypunch Card (Computer Card)*

37

38

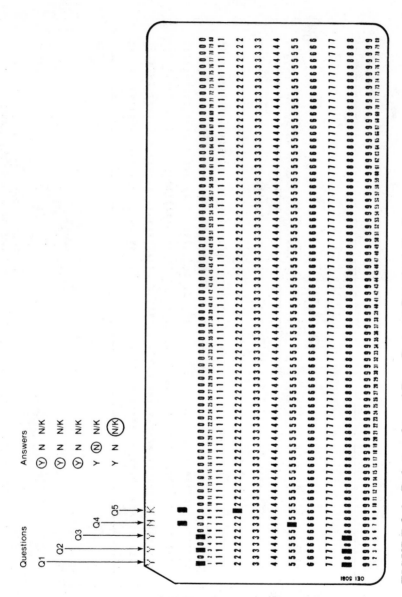

FIGURE 3.6. *Relationship of Raw Data to Punched Card*

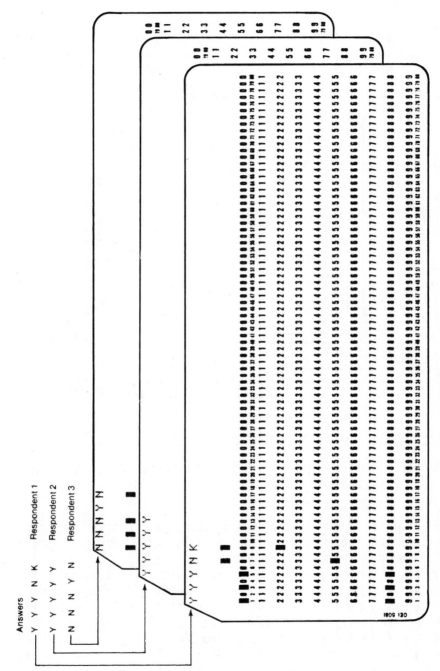

FIGURE 3.7. Relationship of Multiple Respondents, Raw Data, and Punched Cards

39

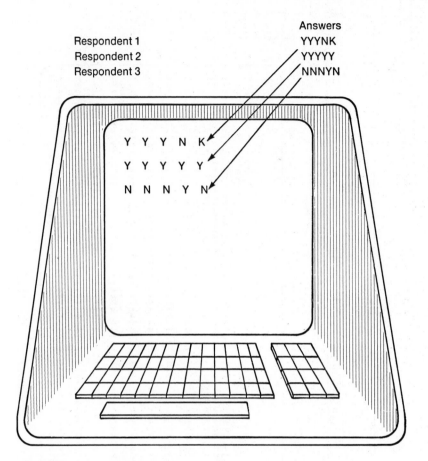

FIGURE 3.8. Relationship of Multiple Respondents, Raw Data, and
            Terminal Lines

line is number 1, the second line is number 2, etc. With data prepared
on cards, only the top card is visible. This difference constitutes a clear
advantage for terminal data preparation.

Punching holes in cards makes a permanent physical record of the
data. Occasionally, first-time terminal users conclude that the televisionlike screen of the terminal is the recording device, and if the screen is
erased, then the information is lost. The screen is roughly the equivalent of the printing at the top of a card. Not printing does not affect
the holes; erasing the screen does not destroy information that has
already been input. All terminals depend on the computer to which

they are connected for recording. Generally this recording happens automatically and is of no concern to the user.

With some idea of good data collection and preparation, the user of a statistical package can confidently expect research efforts to be compatible with the computer. The next step is the actual inputting of the data. The classic procedure—the two-step process of preparing computer cards and then reading the cards into the computer—is the topic of Chapter 5. Direct entry of the data through a terminal is rapidly replacing card preparation. Chapter 4 discusses this new technology.

4

Terminals

Computer terminals are rapidly replacing the card and keypunch combination as the standard computer access method. Pundits have been predicting this for years, but the economics of microprocessors are the direct stimulus for the changeover. Many terminals are in fact specialized computers or general-purpose ones adapted to special terminal functions.

Although not exactly inexpensive, terminals are now in the same price range as many typewriters. This means that they will eventually be commonplace. However, the majority of terminals can do nothing alone. They must be connected to a computer and to a communications system. The number of terminals that a given computer can operate ranges from a half-dozen or so to several hundred. Each active terminal places a demand on the central computer, so such systems can and do become overloaded. Terminal systems have become so popular that it is not uncommon to wait several hours for an available terminal. If the communication is via telephone, the number of available phone lines is often limited and users sometimes must wait a long time for a line.

While it is now possible to have a terminal in every home or office, the communications network and the raw computing power do not exist everywhere. So before you rush out to purchase a terminal, consider carefully whether or not the necessary support is available. Remember that phone lines and computer time are not free.

There is little doubt, however, that terminals have much to offer the user of statistical packages. All of the analysis systems discussed in this book can be run from terminals. With the exception of SCSS and some SAS installations,[1] statistical packages work in basically the same way whether one uses a terminal or a card system. With terminals, significant savings can be realized in data preparation, analysis, reanalysis, and error-checking abilities.

Remember that the essence of a terminal is not the type of machine but rather the type of operating system to which it is linked. Experience with a certain brand of terminal and a particular operating system does not guarantee that one will be able to use the same terminal with a different operating system. This is true even when all of the equipment is from the same manufacturer. Another way to look at this fundamental point is to think of the passivity of a keypunch. Although an instruction is punched onto a card, the computer does not act until the instruction is actually input. In contrast, a terminal is an immediate inputting system. The user is in constant contact with the operating system. In short, terminal and operating system combinations are active and highly dynamic. You, the user, will be communicating directly with the computer, and it is the operating system that makes it all possible.

Sections 4.1 and 4.2 cover types of terminals, their operations, and keyboards. Section 4.3 discusses operating systems, and Section 4.4 offers suggestions on how to cope with a terminal.

4.1   Types of Terminals

Terminals fall into three broad categories: screen (televisionlike) terminals, paper terminals, and special-purpose terminals. Screen types, where a screen and a keyboard are combined, are rapidly becoming the most common. They offer a number of advantages: portability, adaptability to a wide range of communication speeds, ease of corrections (erasures and deletions), ease of file manipulation, and the elimination of mountains of out-dated paper.

Paper terminals, which have no screen, have all the capabilities of the screen types. They substitute typed characters on paper for the lighted screen displays. They have the following advantages: permanent copy (hard copy) is produced; many accept standard fourteen-inch computer paper; and the more expensive models can produce letter-quality typing, so they can be used to type manuscripts. Their major disadvantages are weight, noise, and limited speed.

---

[1] SCSS works only as an interactive terminal system.

Special-purpose terminals, which will rarely be of concern to the social science user, range from those designed for graphics to those that produce paper tape, such as a teletype. In recent years, these special graphics terminals have produced beautiful displays for some movies.

Another major division within the broad area of computer terminals is between IBM systems and ASCII systems. In essence, each system codes the characters in a different way. The same instruction will often be very different between the two systems. (See Section 4.2.1 for a discussion of special keys giving several examples of this difference.)

Terminals are connected to the computer in a variety of ways. The description here will be very general and is summarized in diagram form in Figure 4.1. Direct terminal connection is the simplest. In the case of Figure 4.1, the terminal is located relatively close to the computer and is connected by a cable in much the same way as are card readers, printers, etc. Direct connection allows many convenience features not available at more distant locations. This is a function of both the terminal and the operating system that runs it.

We can see from Figure 4.1 that there are three main types of systems that are not an immediate part of the central computer. The first could be called a distant direct connection. Here the terminals are connected by a special cable that has been installed between them and the computer. Such a system is fast and powerful but also expensive.

A dedicated phone line system operates on telephone lines that are rented to the user 24 hours a day, 7 days a week. This system is moderately fast, but the speed varies with the number of phone lines leased.

The slowest systems are dial-up systems, in which ordinary phone lines—at the standard charge—are used. When such a system is in operation, the receiver of the phone set is placed into the modem to make the connection. (Recall that modem stands for *mo*dulator-*dem*odulator.) Modulated tones are the means of communication. They sound very much like the tones heard in a pushbutton phone system. The same problems that plague the phone system can cause problems for terminal users.

Dial-up systems are further divided into those using acoustic modems and those using high-speed modems. Only the acoustic modem connection gives real portability, since it can be used with any telephone. It is also the slowest—about 30 characters per second. High-speed modems are permanently wired into the phone system for fast transmission. Modern technology has recently provided a new type of modem—the direct-connect modem. This variety uses the modular phone jack and is often built into the terminal itself.

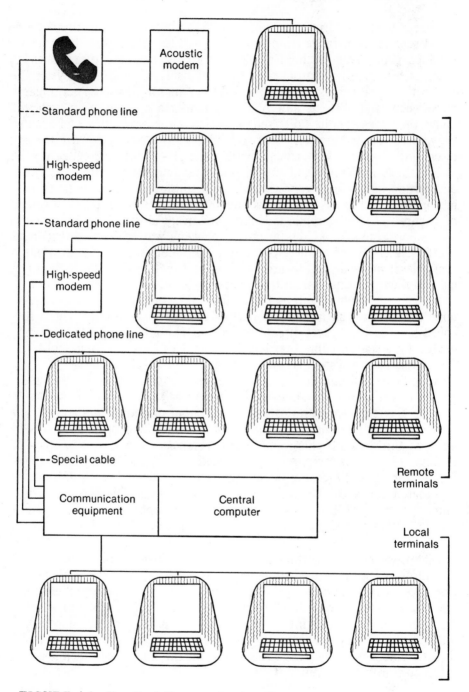

FIGURE 4.1. Terminal Communication Systems

Another way of looking at the differences between terminals is to contrast line terminals with display terminals. Display terminals generally show the user about 20 lines, 10 above and 10 below the line being worked on. Line systems show only the current line and past lines. The major difference is that in line operation only the current line is accurate. The screen is often full, but it does not represent the file. With a display system, you see a portion of the file without the distraction of also viewing the terminal instructions. These display systems usually have some means of scrolling up and down in the file. Note that in both types instructions such as "GO TO LINE 52" do not become a permanent part of the file.

Once the details of the equipment, phone lines, billing, and so forth, have been established, a computer system is simple and efficient. It is impossible to state what a typical social science researcher will have available in the future, but it is clear that the days of cards and keypunches are numbered.

4.2 Operating a Terminal

It is really quite simple to get started on a terminal.

1. The machine must be turned on. The on/off switch might be located on the front, the side, or even on the back in a recess. All screen terminals require a warmup period. "Ready" status is indicated by an indicator light or a clear, unblurred cursor (position marker). Paper terminals do not require a warmup period.
2. With phone systems, the next step is to call up the computer. The computer usually answers with a special tone or signal. Then the handset is inserted into the modem. The modem may have to be turned on by the user, or it may be turned on once each day by the staff. Direct-connect modems require that you issue the command to dial from the keyboard.
3. Paper terminals must be loaded with the correct paper. This procedure is different for each machine. Be sure that the machine is off before attempting to load paper.
4. Some settings may need to be adjusted. All terminals have a variety of switches and settings. One can observe the computer wizards playing with them. The users who are not wizards should leave all internal controls alone. The one exception is the contrast control, which functions similarly to the contrast control for a television. Some terminals also have an intensity control. Set both

controls to comfortable levels, but remember that high settings
may burn the screen.

4.2.1   Keyboards

In Chapter 5, the keyboard of the standard keypunch is discussed in
some detail. To do the same for terminals would require eight or nine
illustrations just to show a representative selection of the available
keyboards. Each manufacturer uses a different set of symbols, so no
comprehensive discussion can be attempted. Instead, the method here
will be to discuss functions and procedures rather than specific keyboard arrangements. This means that when first confronted with a
terminal, you will have to do some exploring and some reading of the
instruction manual.

Terminal users with long experience on standard typewriters are
often at a disadvantage compared to those users who cannot type. This
is because, with the exception of the letter arrangement (QWERTY. . .),
everything varies from machine to machine. When keys are out of the
usual positions, well-learned skills actually slow production. The descriptions and definitions that follow proceed from the familiar to the
unique. The beginner should, of course, read all sections.

   alphabetic keyboard: The standard typewriter arrangement. Gener-
      ally both upper- and lowercase characters are available. Alignment
      of the keys may vary somewhat from the standard typewriter.
      Since the alphabetic keyboard is similar to that of a typewriter,
      one would think that the shift, shiftlock, or uppercase function
      would need no explanation. Problems can arise with many IBM
      terminals, which indicate shift with an open arrow and indicate
      cursor controls with a solid arrow. The saving feature is that shift
      keys are always in the same place as they are on a typewriter. One
      can lock into upper case with a key that is labeled with LOCK,
      ALL CAPS, or an arrow surrounded by a padlock symbol. *Caution:*
      The effect of the lock feature varies; for some terminals, it affects
      only the letters and not the keys that have a different upper- and
      lowercase symbol! A further complication is that some terminals
      unlock the uppercase shift with a lock key, while others unlock it
      with the shift key. A little trial and error will yield the answer.
   numeric keyboard: Two styles are available: a row of numbers above
      the letter keys as on a typewriter, or a ten-digit keyboard arranged
      as on calculators and adding machines. This keyboard is frequently
      separate from and to the right of the main keyboard. Many termi-
      nals incorporate both types.

special characters: Generally a selection of symbols (e.g., #, $, and *) is available, but it does not necessarily include all of the common ones. There is no standard arrangement. Many terminals have both the slash (/) and the backslash (\); they are not the same. One must also distinguish between braces ({ }) and brackets ([ ]).

return, carriage return, CR, enter, or send key: On a typewriter, the carriage return key does just what its name implies. On paper terminals, the printing mechanism returns to the left margin. On screen terminals, the cursor returns to the first column position. Whatever this key is called, it is the way in which the line is actually transmitted to the computer. (Note that systems do exist that react to, or transmit, each character as it is typed and not just to whole lines. Even whole pages can be transmitted at once on some machines.)

cursor control key: Paper terminals do not have cursors, but all screen terminals do. Since, in effect, the screen is a matrix, it is useful to be able to move the cursor around to any position. Generally, this key is marked with an arrow (IBM machines use a solid arrow) pointing in the direction the cursor will move. If the key is depressed continuously, the cursor will move rapidly along the line. Double arrows are for extra-fast movement. The home position is the upper left, and most terminals have a home key. Some terminals have the word HOME written on the key, while others use symbols. IBM uses a small drawing of a screen with an arrow pointing at the home position.

clear key: Most screen terminals have a clear key that removes whatever is present on the screen. This does not mean that your previous work has been lost. Whatever was entered via the enter/ return key is retained in the computer. If you press the clear key in the middle of a line, before pressing the enter or return key, only the current line or part of a line is erased. (Some terminals also have a clear field key that clears only the current line. This is determined by the operating system, so no general rules can be stated.) Some manufacturers place the clear mechanism on a key alternate. In this case, the alternate key (ALT) and the clear key must be depressed at the same time.

reset key: The reset key causes the terminal or the operating system to ignore your most recent instruction. It is not a clear or an erase key. The need for a reset occurs when an impossible or clearly illogical command is issued and the system refuses to act. The reset mechanism allows you to continue. Again, just how and if the reset works depends on the local operation. On some paper terminals, a different form of reset is found that restarts after a mechanical operation, e.g., a new ribbon insertion. You should be aware that terminal reset keys do not have the same function as the reset keys on a computer have.

tab key: Some terminals have tab keys, and some do not. Those that do not may allow specially defined keys to perform the tab command. The word TAB and an arrow pointing to a vertical line seem to be the most popular designators for this key. A backtab is merely a tab from right to left instead of from left to right. The tab on terminals is only roughly like the tab on a typewriter. Terminal tabs and system tabs may be separate and unrelated. You should understand that pressing the tab key generates a tab instruction only; you will not actually see the tab on the screen when the instruction is first typed. On playback of the file, the tab will function. Many popular word processors provide a typewriter-like tab.

space bar: The function of this key is obviously to leave space between characters. Usually this key is the long bar at the bottom of the keyboard, although some terminals have another smaller space bar located on the calculator-style numeric keyboard.

backspace key: This key issues a backspace instruction but (similar to the tab) does not actually backspace on the screen or paper. When the line is printed, it will have been backspaced and underscored (or whatever). The reason for this is that most screen terminals are unable to display multiple characters in a single position. (For a brief discussion of erasing on terminal systems, see Section 4.5.)

break key: Break is a command that interrupts whatever the machine is doing at that moment. Thus, if you need to halt a long printing session, merely push the break key (sometimes labeled ATTN). To resume, press the enter/return key. Note that just exactly where the system will resume depends on the local system.

special keys ESC *and* ~ *(tilde):* There are special-purpose keys on ASCII terminals, ESC (escape) and ~ (called a tilde), which are interchangeable. Both provide the same operation—logical escape. This operation is necessitated by the fact that certain symbols are used as both symbols and commands to the system. For example, if the asterisk (*) key commands double spacing but you want the asterisk itself, a means of escaping from the command function is needed. This is something of an oversimplification, but it illustrates the fact that this situation does arise. And when it does, you should ask the local expert.

special keys â *and* ⌀*:* The two keys â and ⌀ are found only on IBM terminals, although similar functions are often found on other machines. The letter a with a looping line through it is the symbol for delete. It erases whatever is above the cursor, one character at a time. The letter a with the "hat" (circumflex) on it denotes an insert to the left of the cursor. These keys allow for insertions and deletions in lines of text. Remember that the exact nature of these functions is determined by the operating system and not by the terminal itself.

delete key: The ASCII deletion symbol is DEL. Whether or not a key so marked is a legitimate one and precisely how it affects the text must be locally determined.

control keys: Each terminal manufacturer seems bent on providing extra functions as a means of competing with other manufacturers. This results in many specialized keys that are unique to each combination of terminal and operating system. For example, some keys change the form of the cursor or provide reverse video (black on white). Printing terminals have a number of these keys for setting margins and for other printing-related functions. The manuals for each terminal are of some help, but it is usually best to ask a local expert.

special ASCII functions: The ASCII communications standard actually contains 128 separate characters. Obviously upper- and lower-case letters account for 52 of these. Those functions that are not assigned their own key can be obtained as alternates. Either the control key (CTRL) or the escape key (ESC) and another key are depressed simultaneously. For example, CTRL-S means both the control key and the S are depressed. Again, precisely which instruction produces which result is a function of the local system.

4.2.2 Summary

Despite the great variations among machines and despite the extensive local options, terminals are far easier to use than keypunches. Terminal systems are infinitely erasable, whereas once an incorrect hole is punched into a card, the card is effectively ruined. Corrections made on printing terminals can become difficult to read, but otherwise there is no problem in making changes in input.

One must accept the facts that learning to use a terminal entails dealing with certain ambiguities and that errors will be made. On the positive side, terminals and the associated operating systems are much more flexible (and even more forgiving) than the keypunch.

4.3   Operating Systems

The only thing that operating systems have in common is their diversity. IBM no longer dominates the field of operating systems production as it did in the recent past. Indeed, IBM now offers a wide variety of operating systems of its own rather than having only one system available in a variety of models. As minicomputers and microprocessors become widespread, this diversity will increase. Therefore, it is impossible for a text such as this one to describe operating systems in detail. The following will give a broad overview of the subject as well as some helpful hints.

For small computers, there usually is only a single operating system with a simple descriptive name, such as DOS81-2 (disk operating system 1981, version 2). Larger computers may have several operating systems in force: for example, a terminal operating system (TOS), a noninteractive batch operating system (BOS), and an executive operating system (EOS). The batch system might perform all the routine data processing chores, and the terminal system might handle as many as several hundred terminals of different types, while the executive system could be set up to maintain order, keep the books, and allocate resources. Only a small part of the terminal operating system is needed to perform the statistical operations described in this book.

That all operating systems work the same way is a common misconception. Much depends on the nature of the terminal itself and on the communications arrangements and capabilities of the main computer. What works in the engineering department may not work in the business department.

Some operating systems are inherently terminal-oriented, while others may need to be invoked. For the user of a statistics package, the former is generally easier. In a similar vein, most terminal systems are interactive to some degree. This means that the system will engage in a dialogue with the user; e.g., the system might ask you to specify parameters or it might acknowledge your requests.

Most terminal operating systems have fairly specific sign-in procedures. You must tell the system who you are, what account is to be charged, etc. For security reasons, most systems have some form of password. Never choose the obvious (birthdate, initials, or phone number) for your password. In most installations, when you sign in you can expect to define a number of issues, such as how much memory you will need and what resources.

Terminal systems speak to you and often ask specific questions in a language that is terse and filled with acronyms. Remember that spelling is crucial. There are a few likely areas of confusion: when you have requested more memory than you are allowed, when you have filled your present memory allotment, and when you have requested something that does not exist. Another common error is to transmit one command before the system has finished with your previous one. Most systems give some kind of ready symbol, and one should learn to watch for it.

Occasionally, there are noticeable delays between the entry of your request and your receiving the system's response. This is because terminal systems slow down as the amount of traffic increases. There are periods of peak use in most computing environments, and it is wise to try to avoid those times.

Some errors are so common that many systems have special error messages and guidance built in. Some systems even have special libraries of help available on the terminal system—you simply ask for help. In this sense, you will always have a consultant on hand, although it may be a somewhat inflexible one.

On systems that are shared by a wide variety of users (university and college systems, for example), special safety provisions are incorporated that make it next to impossible to "crash" the system by accident. (A crash is the unplanned stopping of computer processing.) Now and then it happens that a program is destroyed, but it is rarely caused by a beginner's mistake. If you are using a terminal that is connected to a minicomputer or microcomputer, the accidental crash is more serious because the smaller systems cannot spare the capacity needed to prevent the trouble. The implications of this depend on the system; but, in general, the smaller the system, the closer the individual user is to the working structure.

4.4   Living with a Terminal

The following list of helpful hints is offered to pass on to beginners the lessons of years of experience. If you are a "pro," you will probably find the list somewhat elementary.

1. If you leave a public terminal for a few minutes, do not turn the contrast or the brightness controls down to zero to save the CRT. The next person might flip the on/off switch several times, attempting to turn the terminal on. Not only will you irritate the next person, but you might also lose your previous work. When it is time to quit, be sure to follow the specific procedure for getting out of the system. Do not just turn it off.

2. A terminal should not be moved while it is turned on. It also should not be operated in a hot, unventilated room. If you find the heat to be unbearable, the machine probably does too. Be careful not to spill coffee or other liquids, as a terminal has more exposed sensitive areas than a television does.
3. Be sure to check paper terminals for correct paper alignment and for a correctly locked paper guide mechanism. If you are thinking of purchasing a paper terminal, remember that they can be quite noisy.
4. If you are contemplating doing manuscript work at a printing terminal, remember that most word-processing systems have a provision for pausing while fresh paper is inserted. Also, special heavyweight-bond continuous paper is available. Most data-processing centers have special machines for trimming and separating continuous paper.
5. You should expect to submit a certain amount of paperwork before being allowed to use a terminal. In a classroom situation, this is likely to be prearranged. The paperwork usually defines who you are as well as who pays and how much. It keeps a record of the computer resources used, and it also defines when you will no longer need the system.

4.5   Special Problems

The ability to erase is a major advantage of terminals over keypunches. The means to accomplish an erasure, however, are not standardized. Since a backspace is a command, backspacing and typing over do not effectively erase material. Many systems function correctly by moving the cursor back and typing over. On systems that do not support cursor controls, you need to use a special erase command—for example, the each sign (@). These systems work on a character-by-character basis. Thus, "NEW YODT@@RK" becomes "NEW YORK" when you play back the line. The each sign erases and is not retained. Display systems and word processors have other procedures for erasing that cannot be detailed here.

Another cursor-related problem occurs when cursor movement is confused with the space insertion ability of the space bar. They do not have the same function. The space bar inserts real blanks in the line, whereas by moving the cursor, one does not insert blanks into the intervening space.

Serious and frustrating problems can occur with terminal systems in which certain keys (functions or characters) are undisplayable or unprintable by your particular terminal. For example, the backspace and tab functions are as real as any others, but they cannot be printed or displayed. This becomes serious in data entry. When such functions are accidentally inserted into a data file, most analysis systems will refuse to process your data. Generally, you will receive a message about bad data from the statistical system. This may cause confusion if when the data are examined they look perfect. If the bad datum is a tab, it cannot be displayed or seen.

There are two solutions to this problem. The first is to observe the displaying of the data. On line terminals, you may observe the cursor moving in a disorderly or nonsequential manner, indicating that it is trying to display the undisplayable. With fast terminals or with those that display 20 or more lines at a time, you will not be able to see the suspicious cursor movement. You can solve this problem by having the data printed on a paper terminal, since most paper terminals are slow. The other solution is to print or display the data in hexadecimal, octal, or whatever code is used by your computer. This allows you to search for the code representing the tab, etc. In practice, your consultant can do this much faster, since he or she is probably familiar with the codes. Note that the data-checking and -cleaning routines provided by the statistical software packages are often not capable of locating this form of error.

One last point: longtime computer users and many programmers seem to prefer using only the uppercase alphabet. This is not necessary; in fact, it is recommended that beginners type their commands in lower case. This allows them to distinguish between what they have said and the computer's responses (presuming that the computer responds in upper case, of course).

5

Keypunches

The various input media were reviewed in Chapters 2 and 3. It was noted that computer cards have been the standard input method up to now but that all-electronic terminal inputting is becoming more popular all the time and will become the standard method in the future. This chapter is devoted to the topic of keypunching cards, since many social science researchers will continue to use this method for some time.

A keypunch is the physical device that punches cards, and, by and large, we will confine our discussion to the IBM 026 and the IBM 029 keypunches. Just as there are varieties of cards and codes, there are also varieties of keypunches. We will discuss the IBM 029 keypunch on the grounds that if you can cope with this device, the others will be relatively straightforward. The analysis packages do not require that any specific punch be used. The IBM models are representative of all common types.

(We can eliminate the IBM 129 keypunch from our discussion because it is clearly a machine for production work. It has far too many controls, options, and complications for reasonable use by the average social scientist. Indeed, your author knows many data-processing specialists who cannot produce a single card with the 129 punch. Do not use that model unless you cannot avoid it.)

The term keypunch refers to the use of keys (as in typewriter keys) to punch holes in computer cards. The holes are organized into a machine-readable code and are physically structured as discussed in Chapter 3.

5.1   IBM 026 and IBM 029 Keypunches

The IBM 026 punch is quite serviceable, but it has one serious draw-back. The code is outdated and is *incorrect* for certain characters. This older model of keypunch produces the BCD (Binary Code Decimal) code, whereas the newer IBM 029 punch produces EBCDIC (Extended Binary Coded Decimal Interchange Code). All of the analysis packages use the more modern EBCDIC code because of its extra functions and capacities. The very considerable overlap between the two coding schemes makes the older machine viable *if* the user is prepared to deal with them carefully. (See Appendix F.)

5.2   General Operation and Mechanical Aspects

We know from the introductory sections that there are varieties of key-punches, codes, and cards; however, for the purpose of instruction, the IBM 029 keypunch will be examined. The physical properties of key-punches are quite similar. That is, they all hold a supply of blank cards, punch them on command, and store the completed cards. Section 5.2.1 discusses the physical layout of the keypunch, Section 5.2.2 deals with the various control switches, and Section 5.2.3 treats the topic of the keyboard.

5.2.1   IBM 029 Keypunch–Physical Layout and Operation

Figure 5.1 gives the location of the major parts of a keypunch. The following discussion is keyed to the numbers in that diagram.

1. *card hopper:* Storage for unpunched cards. The cards are held by a spring-loaded plate with a release lever on the back side. To place cards in the hopper, release the locking lever (push for-ward) and slide the plate to the rear of the punch. Place cards in hopper with the printed side up and the 12-edge on top. Move the plate forward, and release the lever. The cards must be placed squarely and evenly, or the keypunch is likely to jam.
2. *punch station:* The mechanism that makes the holes one column at a time in the cards. You will be unable to see which column is being punched. Column indication is as in item 11 below. The punch station is also the print station. Thus, when a letter key is pressed, both the punch code *and* the printed letter are pro-duced. The keypunch prints on the top edge of the card, column by column.

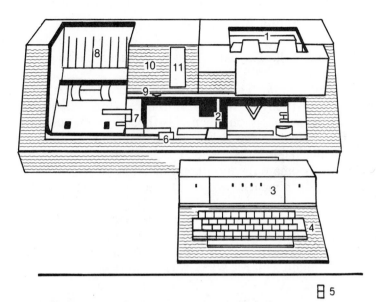

FIGURE 5.1. The IBM 029 Keypunch: (1) Card Hopper, (2) Punching/
Printing Mechanism, or Punch Station, (3) Function Control Switches,
(4) Keyboard, (5) Power, or Main Line, Switch, (6) Backspace Key,
(7) Card Reading Mechanism, or Read Station, (8) Card Stacker, (9)
Program Control Switch (On/Off), (10) Program Control Mechanism,
(11) Column Indicator

3. *function control switches:* The switches that determine the
operations of the punch. These control the major aspects, such
as printing or not printing.
4. *keyboard:* All the keys that produce the letters, numbers,
and special characters as well as control the movement of the
cards. The system is completely electrical and lacks mechanical
connections.
5. *power switch (called the main line switch by IBM):* The on/off
switch. Its location is under the keypunch table (on the right
side). Keypunches consume large amounts of electricity and
should be turned off when not in use.
6. *reverse key:* The backspace key to the rest of the world. It functions just as it does on a typewriter—it does not correct! It allows for skipped columns to be filled and for multiple punches
in the same column.
7. *read station:* The part of the system that allows holes in one
card to be read for the purpose of punching holes on the card in
the punch station. In effect, this duplicates what is read.

8. *card stacker:* The bin for the completed cards. Cards are stacked printed-side up, first card on top. To remove the cards, you simply grasp and pull. The cards may be replaced with reasonable care.

9. *program control switch:* The switch that, in conjunction with the card drum, partially or wholly automates (programs) card movement. This switch is on when it is flipped to the left and off when it is flipped to the right. (See Section 5.5 for details.)

10. *card drum:* A mechanism that, when the program control switch is on, helps to automate the movement of cards. This is done by placing a specially punched card on the drum. See Section 5.5 for details.

11. *column indicator:* A pointer that allows the user to keep track of the column to be punched next. Instead of the linear pointer found on many typewriters, this type of indicator is a rotating wheel located behind a plastic window (for safety).

Cards start at the upper right. They are moved and then are punched or read at the operator's command. Cards are stored at the upper left. They follow a seven-step path. That is, they go from feed hopper, to punch feed, to punch station, to read feed, to read station, to stacker feed, to card stacker.

The punch station and the read station require that the card be registered before punching or reading can begin. That is to say, the card must be placed under a clamping bar that aligns and holds it, which is quite similar to registering a printing plate on a page of newsprint. Registration may be accomplished manually (by pushing the appropriate key) or under program control. The seven-step card path may not be interrupted, since cards cannot be removed until they are placed in the card stacker. [There are exceptions to this rule, but most often attempts to remove cards before they reach the stacker destroy the cards. In addition, physical damage to the mechanism may result, so remember to move cards to the stacker before you try to take them out (see Section 5.4).]

The descriptions above may sound as if too much emphasis has been placed on the mechanical workings of a keypunch. The purpose is to lessen beginners' anxieties and thereby benefit their performance. As you will discover, general-use keypunches are much abused and are often in need of repair. Why? Because most people who use them know virtually nothing about them other than that they have an obvious typewriterlike function.

5.2.2 Control Switches

The power switch and the program control switch have been discussed in Section 5.2.1. The remaining control switches are located in a row just above the keyboard itself. Figure 5.2 shows the exact locations of these switches. What follows is a list of these switches and their functions. The numbers correspond to those shown in Figure 5.2.

1. *clear:* This switch removes all cards to the stacker. It does not affect those in the feed hopper, however. This is a momentary switch; i.e., it has a spring that will always turn it off after use. Its principal function is to move any cards left behind by the preceding user (if he or she has been so inconsiderate as to leave some cards behind for you to deal with).
2. *left zero (LZ) print:* This switch automatically suppresses all left, or leading, zeros. Turn the switch to on, and all the zeros are printed. Blanks are not affected, nor is the punching of zeros. Only the printing is involved here. For easier proofreading, it is recommended that you set this switch to the on position.
3. *print:* When this switch is in the off position, the card will be punched without printing along the top edge. It is difficult to read holes, so you should always keep this switch in the on position.

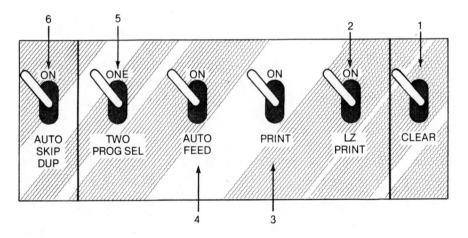

FIGURE 5.2. Control Switches

4. *automatic card feed (AUTO FEED):* This switch moves cards from the feed bin to the punch feed station and moves any card in the punch feed station to the punch feed itself. Thus, the punch can save the user the effort of manually readying a new card. Of course, it does this procedure only when the current card is finished. To start this process, the user must move *two* cards from the feed bin. Then after each card is finished, a new card is registered and another card is moved to the feed station.

5. *program select (PROG SEL) (ONE or TWO):* As previously noted, movement can be automated with the memory drum and the program control switch. There are actually two different programs available. The program select switch selects between the two. There is no off position; for off, use the program control switch. (See Section 5.5 on program control.)

6. *automatic skipping (AUTO SKIP) and duplicating (AUTO DUP):* This switch, in conjunction with the memory drum, provides control over the automatic functions of skipping and of duplicating. In effect, this is a way to regain manual control. (See Section 5.5.)

As a general rule, it pays to set all of these switches immediately after turning the power on. The most critical one is the print switch, since it is very difficult to check for errors without seeing the information in print.

5.3   Keyboard Details

Although keyboards may vary in their layout, they all serve the same purpose: to encode letters, numbers, and special characters. Card movement is also determined by special control keys, which is discussed in Section 5.4. In this section, we will discuss the alphabetic and numeric keys, the special characters, and the periods, commas, and quotation marks. Also included are special problems associated with the IBM 029 keyboard. (*Note:* Sections 5.3.1 through 5.3.4 assume manual control of card movement. Automated control is discussed in Section 5.5.)

5.3.1   The Alphabetic Keys

Obviously, the keys with letters on them cause those same letters (alphabetic characters) to be punched. The letters are arranged in the

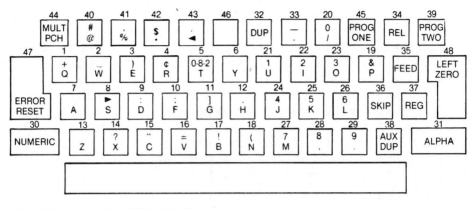

FIGURE 5.3. The IBM 029 Keyboard

standard typewriter pattern, also called the QWERTY keyboard (after the first six letters). With certain exceptions, computers process only capital, or uppercase, letters. The keypunch provides uppercase letters *without* having to shift. Thus, looking at the key in the uppermost left position (key 1) in Figure 5.3, we see that it is the alphabetic character Q and the special character + (plus sign). The Q is the unshifted result; the plus sign results only by shifting to upper case. Shifting is discussed in Sections 5.3.2 and 5.4. The remainder of the keyboard is in standard typewriter arrangement. Note that the period and the comma are also in the QWERTY standard positions (unshifted). The letter I cannot be used for the number 1, or vice versa. Nor can the letter l be used for the number 1 as it can on a typewriter.

5.3.2 The Numeric Keys

The numeric keys are, as Figure 5.3 illustrates, ten keys toward the right of the keypunch keyboard (keys 20 through 29). They are upper case and require a shift. Numeric shifting is accomplished by depressing the key marked NUMERIC in the extreme lower left-hand corner (key 30). This key must be depressed continuously; there is no lock as there is on a typewriter. There is, of course, a way to produce all uppercase characters, but it can operate only when the functions are under automatic control.

The location of the 0 (zero), above the 3, is usually not a significant problem. However, the pattern of the numeric keys does often prove to be a problem. From Figure 5.3, it can be seen that the pattern is the opposite of that found on electronic calculators and adding machines. The top row on calculators is 7, 8, 9 and not 1, 2, 3. The result of this difference, unfortunately, is often frustration and unnecessary errors. Some newer keypunches have a separate numeric key pad arranged in the calculator pattern.

The most common errors when keypunching numbers are the result of incorrect numeric shifting, which produces letters instead of numbers and numbers instead of letters. Note also that the period (decimal place) is lower case and shares the same key with the number 9. To avoid shifting for decimal places, you can use the uppercase decimal place, key 43 in Figure 5.3. Of course, the results in either case are identical.

5.3.3   The Special-Character Keys

The special-character keys are as indicated on the keys in Figure 5.3. Most of them are familiar enough that it is unnecessary to note, say, that the key marked + is the plus sign. However, while most of the special characters are upper case, the @, %, *, <, /, and - are all lower case. Note also that the minus sign (key 33) is the same in upper and lower case. The minus sign and the hyphen are actually the same character.

It is possible to confuse the special characters; for example, the uppercase W (key 2) is an underscore and not a minus sign. Uppercase G (key 11) is neither a minus sign nor an underscore but rather the logical NOT.[1] Uppercase Y (key 6) is the vertical bar, or logical OR; and uppercase B (key 17) is the exclamation mark. Lastly, uppercase T (key 5) produces the 0-8-2 (or 0-2-8) code. This key does not have a name, nor will a symbol be printed on the card. For further discussion of the 0-8-2 code, see Section 5.7. (Some IBM 029 keypunches have a 1-5-2 key, but in fifteen years of computer work, this author has never used that code.)

5.3.4   Special Problems with Parentheses, Commas, and Quotation Marks

A common keypunching error is to confuse the commas (keys 28 and 41) with the apostrophe, or prime (key 12). Occasionally, quotation marks (key 15) are substituted for the apostrophe. When keypunching information requires these symbols, be sure to determine exactly which mark is needed. Substitutions for the comma in format statements (see

---

[1] The logical NOT key indicates that something "is not equal to," the $\neq$ sign of mathematics.

Chapter 6) are particularly troublesome for beginners. There are two identical commas, one in ALPHA and one in NUMERIC shift.

Left and right parentheses are separate codes and must be distinguished, as on a typewriter. Generally, a left parenthesis will be followed by a right, and, in the case of multiple parentheses, the number of right parentheses will equal the number of left parentheses.

The same balancing rule is true for opening and closing quotation marks. Things must be balanced. This is not a function of any cosmic symbolism but is simply because the analysis packages are constructed according to logical systems that work one way only.

5.3.5   The Space Bar

The space bar produces blanks as required and has repetition capabilities. Simply press the bar until the column indicator points to the correct column.

5.4   Keypunch Control and Control Keys

In this section, manual control of the punch will be discussed. Broadly speaking, the control keys can be divided into keys affecting card movement, keys affecting what is currently being punched, and rarely used keys affecting specialized functions. By movement we mean the progression from the card hopper to the card stacker. The key numbers refer to those shown in Figure 5.3.

5.4.1   Movement Control Keys: REL (Release), FEED, and REG (Register)

After meeting the first requirement for punching (which is the movement of a card from the card hopper to the punch feed station), the user begins by depressing the feed (FEED) key (key 35). When this key is depressed again, another card will move from the bin and the first card will move to the punch station. This is the correct procedure when the AUTO FEED switch is on. Continuous pressure on the feed key will result in jamming the machine. When the AUTO FEED switch is off, the card must be placed in the punching mechanism by keying an instruction.

The card is registered in the punching mechanism by depressing the register (REG) key (key 37). Continuous depression of the register key will cause the keyboard to lock. The board can be reset with the key marked ERROR RESET (discussed below). Once a card has been registered, you can depress any key other than a control key to cause a punch to be made (in the column indicated by the column indicator).

The card in the punch mechanism is moved to the read station by pressing the release (REL) key (key 34). Once punches have been made in the last column (column 80), the machine automatically moves the card to the read station; the release key does not have to be pressed.

Once in the read station, the card must again be registered with the register key. Of course, this time the registration takes place in the read station. Removal of the card is accomplished at any point by depressing the release key. Under manual control, the card will not be placed in the card stacker. It is necessary to depress the register key or the feed key for a new card.

These three keys are the basics of punch control. Novices should be sure that they understand these keys. Theoretically, it is possible to skip the rest of this chapter, which deals with the more sophisticated aspects of card control. However, the dividends to be gained from familiarity with the later sections will be realized when you are punching more than a few cards.

5.4.2 Punch Control Keys: NUMERIC, ERROR RESET, MULT PCH (Multiple Punch), and DUP (Duplicate)

There are three important keys that control the punch itself. In the lower left-hand corner of the keyboard the shift key marked NUMERIC (key 30), which is slightly larger than an alphabetic key, shifts to the upper case. When the keyboard is locked into upper case (see next section), the NUMERIC shift has no effect.

The key marked ERROR RESET (key 47), located just above the NUMERIC shift, releases the keyboard in case of an error. What this usually means is that the operator has initiated an incorrect sequence of steps. For example, the operator may have punched or released before registering. If the punch remains locked, the user is obviously doing something wrong, and should carefully check the procedures.

The key marked MULT PCH (key 44), located just above the ERROR RESET key, is for multiple punching. This function allows the user to create punch codes that are not available on the keyboard. The card being read or punched will not advance as long as this key is depressed. In practice, this occurs most often when a keypunch with a 48-character set is used to produce a 64-character set. As was mentioned previously, this is the case when using an IBM 026 punch.

The duplicating key, marked DUP, is an extremely useful punch control key. By depressing the duplicating key (key 32), one produces an identical copy of the punch code from the card in the read station onto the card in the punch station. This can save the user much effort.

Cards need not be completely repunched—merely copied. For example, say that you are punching this complex control card:

FORMAT   FIXED(10X,2F9.4,3(2F3.1,1X)F2.0/10X,7F5.0)

Following the last zero, you want to punch a right parenthesis, but say you inadvertently fail to shift to numeric (upper case), thereby punching an E instead. You could start from scratch and perhaps make another error, or you could simply release the card to the read feed station. While it is at that station, you should note in which column of the card the error is located (57 in this example). The next steps are to feed in a fresh card, register it, and then to duplicate the first card through column 56 by depressing the duplicating key. (Recall that the column to be punched is indicated by the pointer above the read feed station.) Duplication will continue column by column until the duplicating key is released. A little practice will show you that the key should be released before reaching the column to be corrected. Depressing the key briefly allows a single-column duplication. Of course, the right parenthesis must be punched (after shifting to numeric) when the mechanism reaches column 57.

Errors in any column can be corrected in this way. In many instances, however, errors are detected after the card has been removed from the keypunch. In this case, the card in error may be inserted into the read feed station for duplication. A careful examination of the read feed station will reveal two notches in the plastic guard; one notch is for the top edge (the 12 edge), and the other is for the bottom (the 9 edge) of the card. The card is now placed in the station by sliding it, with the column-1 edge first, into the notches. It will protrude slightly. It may be registered at this point, or a fresh card may be fed in and registered, which will also register the error card. Duplication then proceeds as above.

5.4.3   Production Keys: PROG ONE (Program One), PROG TWO (Program Two), LEFT ZERO, MC (Master Card), AUX DUP (Auxiliary Duplication), ALPHA (Alphabetic), and SKIP

The program one (PROG ONE) key (key 45) and the program two (PROG TWO) key (key 37) provide for switching from one automatic control program to the other. Of course, if the program control switch is off, both keys are nonfunctional.

The key marked LEFT ZERO (key 48) starts printing zeros at the left when the punch has been specially programmed to do so. This key functions only on IBM 029 Model B punches. For the social scientist, this is a superfluous key.

The MC (master card) key (key 46) and the AUX DUP (auxiliary duplication) key (key 38) are also little used keys. Indeed, on most IBM 029 keypunches these keys are not even connected! They may be safely ignored.

The key marked ALPHA (key 31) has the opposite function of the NUMERIC shift and allows for shifting to the lower case from the upper case. Recall that locking the punch into upper case requires automatic control.

Finally, the key labeled SKIP (key 36) allows selective, semiautomatic skipping of portions of cards. As we will see in the next section, a card is divided into fields under automatic control. The skip key allows one the option of skipping to the next field. Thus, in a given problem, if information sometimes is punched in columns 1 to 15 and sometimes is not, the user may skip to column 15 simply by depressing the skip key. This presumes that the program control switch is on and that a proper program card was prepared.

5.5   Automatic Control—The Program Drum

Even a novice will soon outgrow the limited capabilities of the keypunch that have been presented thus far. The next step is to use the memory abilities of the punch, i.e., the ability to execute certain actions at specific points in the production of one or more cards. A simple case in point is data punching of numeric characters. It is annoying to have to hold the NUMERIC key in the depressed position, and it makes errors more likely to occur. It also prevents the fingers of the left hand from maintaining the position on the raw data forms. A totally blank card placed on the memory drum automatically locks the punch into upper case.

This example illustrates how a modest increase in keypunching sophistication can lead to large gains in convenience and speed. The following sections present the mechanical aspects, control card codes, and some examples of automatic control systems for the IBM 029 punch.

5.5.1   Mechanical Aspects

The program drum is located on top of the column position indicator (see number 11 in Figure 5.1). To remove or to replace the drum, you must set the program control switch in the off position and have the

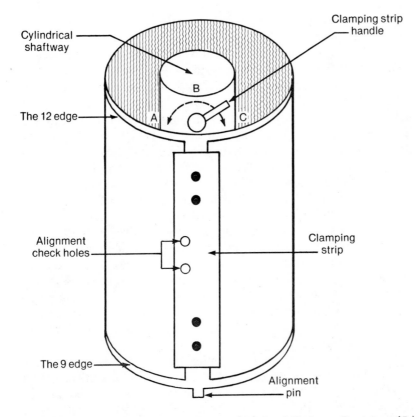

Cylindrical shaftway

Clamping strip handle

The 12 edge

Alignment check holes

Clamping strip

The 9 edge

Alignment pin

FIGURE 5.4. The Program Drum: (A) Card Release Position, (B) Clamp Column-80 End Position, (C) Clamp Column-1 End Position

hinged cover rotated forward, exposing the drum. Figure 5.4 illustrates the basic features of the drum, which is a metal cylinder with a cylindrical shaftway in the center. On the bottom edge is an alignment pin that fits into a hole in the column indicator. The drum is placed into the keypunch by sliding the central shaftway over the top of the post that protrudes from the top of the column indicator. Then the alignment pin is registered with its hole. When the drum is pressed down slightly, it should click into position.

After the drum is in place, the cover can be closed, but *do not* turn the program switch on! The program reading mechanism can be damaged by operation without a program card in place on the drum. With the drum cover closed, check to see that the drum is correctly in place by pressing the release key. The drum should rotate as the column index rotates (one complete cycle). *Warning: Do not operate the keypunch with the drum cover open!*

To remove the drum, turn the program switch to the off position (to the right) and turn the column indicator to 1. The drum cannot be removed or inserted except in this position. Next, open the cover and slide the drum upward off the shaft.

If you want to place a card on the drum, remove the drum as above, and (referring to Figure 5.4) locate the bright silver strip on the outside of the drum just above the alignment pin. This is the clamping strip. It holds the card on the drum with the printed side out. Move the clamping strip handle to position *A*. This is the card-release position. Place the column-80 end of the card under the left-hand edge of the clamping strip. Holding the card in position, move the clamping strip handle to the *B* position. The clamp should hold the card firmly, as the card must be wrapped around the drum. Also, you should be able to see the edge of the card in the alignment check holes. The card edge should evenly bisect the holes. Next, wrap the card around the drum, placing the column-1 end *under* the right side of the clamping strip. Holding the card in place, move the clamping strip handle to position *C*. The card should now be held firmly at both ends and should not be loose or out of alignment.

If the preceding directions seem complex or forbidding, relax. After a few tries, it will be like tying your shoes. This simple operation allows the user to lock the punch into all-numeric operation with a blank card. Further automation can be accomplished by punching special codes into a program card. (*Note:* The program drum will click into its proper position only when the clamping handle is in position *C*.)

Note that the keypunch you want to use may not have a drum in it. As with anything portable, people keep them in their desks, brief cases, etc. The drums are standardized and can be moved from one machine to another. Also note that the program drum cards can be saved and used again. Your consultant will be able to help you locate a drum.

5.5.2   Program Card Operation Codes

The term code is misleading when applied to the program codes. Keypunch control instructions would be better, although this term is more cumbersome. The various codes instruct the punch in much the same way that a programmer instructs the computer. In fact, we have already learned the first instruction—shift to numeric is the message carried by a blank card on the drum. Recall that the operator can override this instruction by depressing the alphabetic shift key.

When a drum loaded with a set of instructions is inserted and the program control switch is on, a series of small star-shaped wheels is pressed down on the drum-and-card combination. These wheels allow the punch to read your instructions. The organization of this instructional system is on a column and column-group basis. Groups are called fields, and naturally they can be from 1 to 80 columns in length. Basically, there are five instructions possible: field definition, automatic skip, automatic duplication, alphabetic shift, and numeric shift. The functions are listed in Table 5.1, along with the codes that produce them.

Thus, it can readily be seen that a card containing 80 blanks instructs the punch to shift to numeric 80 times. A somewhat more elegant means of accomplishing this is to use the field definition. One cannot begin with a field definition, but one can continue whatever has been started. The absence of a 12-zone punch defines the end of a field. Our all-numeric card would thus become this:

ϕ&&&&&&&&&&&&&&&&&&&&&&&&&&&&&&&&&&&&&&&&&&&&&&&&&&&&&&&&&&&&&&&&&&&&&&&&&&&&&&&&&&&&

If the last 20 columns were to be blank, we would have:

ϕ&&&&&&&&&&&&&&&&&&&&&&&&&&&&&&&&&&&&&&&&&&&&&&&&&&&&&&&&&&&&-&&&&&&&&&&&&&&&&&&&&&&

Remember that it is the absence of a 12-zone punch that defines the end of a field. In the above example, the first field ends with column 59 and the second ends with column 80.

Consider a situation in which numeric information will be punched in columns 1–15, nothing will be punched in columns 16–36, letters will be punched in columns 37–40, and nothing will be punched in the remaining columns. The control card would take this form:

ϕ&&&&&&&&&&&&&&-&&&&&&&&&&&&&&&&&&&&&1111-&&&&&&&&&&&&&&&&&&&&&&&&&&&&&&&&&&&&&&&&&&

In this example, one would think that the alphabetic shift (code 1) would be continued with the & character. We immediately have an

TABLE 5.1. *Program One Codes (for IBM 029 Keypunch)*

Function	Code (punch)	Character
Field definition	12	&
Start automatic skip	11	-
Start automatic duplication	0 (zero)	0
Alphabetic shift	1 (one)	1
Numeric shift (the normal, or unshifted position)	ϕ	none

exception. The normal mode is numeric with the control switch on. Hence, it is necessary to continuously operate the alphabetic shift. This card would cause the following sequence of actions:

1. Start with numeric shift in column 1.
2. Continue with numeric shift through column 15.
3. Begin with automatic skip in column 16.
4. Continue with automatic skip through column 36.
5. Begin with alphabetic shift in column 37.
6. Continue with alphabetic shift through column 40.
7. Begin with automatic skip in column 41.
8. Continue with automatic skip to end of card.

We are assuming in the above sequence that the automatic duplication/skip switch is on.

The last feature to be demonstrated is automatic duplicating. If, in the above example, the researcher thought to code the cards for the particular study and for the principal investigator, a reasonable strategy would be to punch this information at the end of every card. (A study in thermal biofeedback by your author was coded TEMP1-DMK.) Nine characters can be automatically punched in the last columns by altering the above control card as follows:

Ⅾ&&&&&&&&&&&&&&-&&&&&&&&&&&&&&&&&&&&1111-&&&&&&&&&&&&&&&&&&&&&&&&&&&&&&0&&&&&&&&&

Automatic duplication can function only if there is something to duplicate. The information must be manually punched onto the first data card. Thereafter, whatever appears in columns 72–80 will be duplicated. Note that if there is nothing to duplicate, the punch will lock and error reset will not unlock it. In this case, turning the drum switch to off will unlock the punch.

Another useful feature is the skip key. When using the memory drum, you will cause the punch to skip until the end of the field, one field at a time, by pressing this key. Any field can be skipped in this way provided that the field is defined with ampersands. There is an important exception to the use of the skip key when the field is defined by repeated instructions and not by field definitions. This is the case when alphabetic shifts are defined by repeated ones: a 1-column skip is produced by depressing the skip key. For example, if the first 15 columns are alphabetic and the second 15 are numeric (recall that the alphabetic shift cannot be continued by using the field-definition procedure), we have:

111111111111111 &&&&&&&&&&&&&&

After a short alphabetic section is punched (the last name Lum, for example), it is not possible to skip to column 16 without pushing the skip key 12 times or the space bar an equal number of times. The solution is to code both the 1 (alphabetic shift) and the &. This may be accomplished in two ways. One way is multiple punching: punch both a one and an ampersand in each column after the first. The second method, since an ampersand is a 12-punch and a one is a 1-punch, is to combine the two to yield the code for the letter A (12-1). Thus, for the above example, we have this:

1AAAAAAAAAAAAAA &&&&&&&&&&&&&&

If both fields were alphabetic, then we would have:

1AAAAAAAAAAAAAA1AAAAAAAAAAAAAA

In effect, two fields are created, which allows multicolumn skipping at your option.

Automatic skip and automatic duplication function in the same way but do not require a combination code. If we need a series of cards on which the first 10 columns are to be skipped, the next 15 are to be alphabetic, the next 13 are to be duplicated, and we are then to skip to the next card, we would prepare a drum card as follows:

−&&&&&&&&&1AAAAAAAAAAAAAAA0&&&&&&&&&&&&−&&&&&&&&&&&&&&&&&&&&&&&&&&&&&&&&&&&&&&&&&&&&&&

Note that this gives a combination of automatic and manual control of the keypunch.

Before advancing to applied keypunch problems, we should note that 1-column fields *are* permitted. Indeed, they are quite useful. For example, if data have been punched in accordance with the guidelines presented in the earlier chapters (blanks for readability), and if the punch skips too soon or fails to skip at all, you will know that something is wrong. In effect, you can partially proofread as you go. This can be a significant advantage.

Thus, two completely different sets of instructions can be used (alternately, of course). Program selection is under the control of the program select switch and the PROG ONE and PROG TWO keys.

In circumstances where it is necessary to rapidly change the punching instructions, PROG TWO codes may be used. These codes are all numeric and do not overlap the PROG ONE codes. They are shown in Table 5.2.

TABLE 5.2. Program Two Codes
(for IBM 029 Keypunch)

Function	Code (punch)	Character
Field definition	4	4
Start auto skip	5	5
Start auto dup	6	6
Alpha shift	7	7
Numeric	⌀	none

5.6   Keypunch Automation Applied

Though many of you already have plunged ahead and begun work with the keypunch, one task remains: practice, practice, practice in automating the keypunch. In learning to do so, you will not only save time and effort, but you will also get some "feel" for the process of programming in general. Most programming languages require the same restrictive, linear, and logical approach that is part of preparing a program card for automatic keypunching.

Your solutions should be very similar to those given in the exercises that follow; however, remember that there is usually a variety of acceptable answers. A close comparison of your solution with the text answer will help you to learn the little tricks of keypunch automation.

═══════ **EXERCISE 1** ═══════════════════════════════════

Write a program control card for the following data card.

1794 M 147321   134 24 1 2 3 2 1 2 2 9

  Answer:

⌀&&&-1-&&&&&-& &&-&- - - - - - - - - -&&&&&&&&&&&&&&&&&&&&&&&&&&&&&&&&&&&&&&&&&&&&

This example is fairly typical of the data from a short survey consisting of answers to questions about the respondent's identification (ID), the respondent's gender, demographic information, geographical location (a number), and the coded answers to eight other questions. See how the skip instruction between each question was used. Also note the use of the automatic skip function to remove the card once the last question has been punched.

If there were a follow-up survey to that of Exercise 1, consisting of six questions, the additional data could be added by placing the old cards in the feed hopper and skipping the old information. There are more sophisticated ways to accomplish this, but few are as direct or as effective.

═══════ **EXERCISE 2** ═══

Write a program control card for the following data card. Maximize speed.

	Old data	New data

1794 M 147321   134 24   1 2 3 2 1 2 2 9   6 7 5 4 9 9

Answer:

–&&&&&&&&&&&&&&&&&&&&&&&&&&&&&&&&&&&–& – – – – – –&&&&&&&&&&&&&&&&&&&&&&&&&&&

The six new coded answers could be located in any of the available columns. In other words, there is nothing sacred about the text solution. The same can be said about skipping columns between questions.

Practice with automatic duplication may be gained by reconsidering the above survey under the assumption that only men (M) were questioned and all were from the same geographical location (134).

═══════ **EXERCISE 3** ═══

Write a program card as in Exercise 2, modified to autoduplicate gender (M) and location (134).

Answer:

b&&&–0– &&&&&–&0&&– &–& – – – – – – – –& – – – – – –&&&&&&&&&&&&&&&&&&&&&&&&&&&

A variation on the text answer would be to duplicate the blanks adjacent to gender and location. The net effect would be the same.

To demonstrate the use of the alphabetic shift, consider a study in developmental psychology wherein the same group of children were studied repeatedly. In such studies, it is often a great convenience to consider each subject by name and not by an arbitrary ID number. Thomas Sampson is much easier to remember than number 72. Also, the first 21 spaces on the data card are to be reserved for the name, followed by the age, gender, etc.

========= EXERCISE 4 =====================================

Write a program card for the following data card.

SAMPSON, THOMAS   72 M (etc. . . .)

     Answer:

1AAAAAAAAAAAAAAAAAAAAA- -1- (etc. . . .)

11111111111111111111- -1- (etc. . . .)

The second solution given is a poor choice because it does not allow the operator to skip to column 22 when the name is finished.

Another common need when keypunching social science data is the production of repetitive information. In Exercise 4, if the subjects were tested every three months for five years, the researcher would certainly wish to record each testing date. This could be done with a code (37 = 14 March 78, for example) or with the actual date. If all subjects are tested on the same date, the automatic duplication instruction will reproduce the date on all cards.[2] Remember that the first date must be punched manually.

========= EXERCISE 5 =====================================

Write a program control card for the following data card, where the date 14 March 78 is in the last six columns.

SAMPSON, THOMAS   72 M (etc. . . .)

     Answer:

1AAAAAAAAAAAAAAAAAAAAA- -1- (etc. . . .)   0&&&&&

----

[2] *Note:* Dates are always punched in the order year, month, and day.

Of course, the use of automatic duplication is not confined to the end of a card. When punching control cards for the SPSS analysis package, you may find that the following program control card is convenient:

1AAAAAAAAAAAAAA1AAAAAAAAAAAAAAAAAAAAAAAAAAAAAAAAAAAAAAAAAAAAAAAAAAAAAAAAAAAAAAAAAAAAA

Remember that the use of this card requires shifting to numeric with the numeric shift key. The card's usefulness lies in its ability to skip to the correct place on command. This might appear to be a minor advantage in that, after the control cards have been punched, this card is awkward to use for data cards. A simple solution exists. When punching data, switch to program two (the PROG TWO switch or key). Under program two control, this program control card contains only blanks—the all-numeric code for data punching.

The last exercise is a final review. You should find yourself using one of every character.

===== EXERCISE 6 ===

Write a program control card for the following data, on the basis that they are part of an experimental study of the sexual activity of several groups of laboratory rats. As a precautionary measure, punch your initials in the last three columns of each card.

001 1 M 17

Answer:

- -1-   -ƀƀƀ&&&&&&&&&&&&&&&&&&&&&&&&&&&&&&&&&&&&&&&&&&&&&&&&&&&&&&&&&&&&&&&&&&&&&&&&&0&&

Your solution for Exercise 6 should include all of the instruction codes in Table 5.1. Once a program card has been prepared for a study such as the one in this exercise, you can punch the cards almost as rapidly as you could enter the figures into a calculator.

At this point, you should feel comfortable in instructing the punch to skip, duplicate, stop, start, and shift. This implies that when a program card is used, the operator has no control over card movement unless the program control switch is turned off. This is not so—cards may be fed (FEED), registered (REG), and released (REL) by placing the automatic duplication/skip and automatic feed switches in the off position. Despite the instruction on the drum card, the operator can regain control at any time by this technique. Shift instructions continue to be in force, however.

Having worked through the foregoing exercises, you probably feel that you know a lot about the keypunch itself, something about automating it, but relatively little about producing a deck of cards. Appendix G reproduces data from an experiment in physiological psychology. Those of you who are political scientists or sociologists and are therefore not very interested in this type of research should nevertheless take the trouble to punch the data, as called for in Exercise 7. You might pretend that the study is about humans and drug abuse. "Weight at trial 1" can become the number of years the subject has been substance-free, and "milliliters morphine" can become the amount of an anti-addiction drug used. In this case, the researcher would be interested in the relationship between the dosage and the time free from addiction. For the present, it is important simply to get some practice punching data.

═════ EXERCISE 7 ══

Prepare a program control card for the morphine-stress data in Appendix G. The arrangement of the data can be whatever you think is best. It is advisable but not necessary to skip spaces to facilitate proofreading. Then you should actually punch all 40 of these cards, which will probably take about 30 minutes to complete.

These cards should be saved because they will be referred to in several later chapters. Although using your own data for practice is okay, you should still have the cards checked by an experienced person.

If you have mastered the above, you are a member of a select group— i.e., social scientists who can use most of the functions of the keypunch. Indeed, the majority of all keypunch users simply bash their way through the operations under manual control. This method works, but it lacks understanding and style, and it also requires unnecessary effort. Remember that there will always be a variety of solutions and little reason to say that one is better than another.

5.7   Common Errors and Problems

Keypunches are the most used and abused of all computer-related equipment. Many individuals misuse keypunches and then leave the machine for someone else to fix. In school settings, this problem gets worse as semester deadlines approach. Many problems can be fixed only by service technicians. However, recognition of the problems will help the user to avoid a malfunctioning punch and to keep the machines in better condition over the long run. This section should acquaint you with the more common problems of using keypunches and with their solutions.

The most common problem is that of a missing memory drum. Manual operation is possible, but it is certainly an inconvenience. If drums are not available, ask around. Individuals sometimes use a machine on manual after taking the drum off. Borrow it! Consistent problems with missing drums mean that the computer center is negligent. A little gentle complaining is in order. Center personnel quite often simply forget to put the drums back. For that reason, some centers issue drums by request only.

A missing memory drum is a minor problem compared to a missing star wheel. These wheels allow the punch to read the instruction on the program control card. You can locate the wheels by inserting a loaded memory drum and then turning the drum control switch on. You should see eight arms through the plastic window. Each arm is tipped with a star wheel. When these arms contact the memory drum, the punch reads your instructions column by column. For unknown mechanical reasons, these wheels are not very secure in their sockets. Rough or improper insertion of the drum easily dislodges them. Thus, it is prudent to check the punch before use. Obviously, if the star wheels are missing, the instructions on the program control card are useless.

There is also the possibility that only half of the star wheels are missing. You will note that they are arranged in two groups of four. Each wheel assembly reads one row on the card. If only the first four wheels are missing, then program two instructions will function properly. The reverse is also true. For example, if the wheel for the row of 1's is missing, only the alphabetic shift instruction will be ineffective. Numeric, automatic skip, and automatic duplication will still function. Thus, despite missing wheels, the memory drum system might meet your needs of the moment. *Note:* Star wheels are quite easy to replace, but the amateur repairperson (i.e., the social scientist) probably should not attempt it.

Another major area of punch problems is the wide variety of card jams caused by mechanical misalignment. Most jams occur as the card enters the punching station. Do not remove the card by pulling it out of the punch station, as this practice eventually causes deterioration of the rollers that advance the cards. The best way to clear a jam is to press the release key (REL). If the release key fails, it may be that the card never completely entered the punch mechanism or that the card is out of alignment to the extent that it cannot pass the punch. In this case, the punch may be cleared by gently sliding the card to the right.

In the situation where the card is in the punching mechanism, the pressure-roll release lever is used. This lever is below and to the left of the program drum. To clear a card, raise the cover, press the release lever, and carefully slide the card to the left or right. If torn pieces remain in the punch, remove them by pushing them with the edge of a scrap card. *Do not* insert anything else into the punch! The read station may be cleared in a similar way. After clearing the jam, you should check the cards in the feed hopper for alignment. This is a frequent problem spot. Further adjustment of the punch to prevent jamming is a job for professional service people. Place an out-of-order sign on the punch, and move to another machine.

Another common problem is the misalignment of the card punches; i.e., the holes are not quite in the proper places. Card readers are less tolerant of misalignment than are keypunches. Misalignment can be caused by mechanical problems or by sloppy procedures. Misaligned cards need not be replaced manually; you may duplicate them in the usual way. Keypunches are tolerant of misuse. If the punch locks while duplicating the misaligned card, clear that card and begin to punch manually. A lockup indicates that the tolerance of the punch has been exceeded.

Keyboard lockup will occur whenever an unacceptable code is encountered. For example, the code 7-8-9 is theoretically possible, but it means nothing to the machine and is not used. The lockup feature allows bad codes to be detected immediately.

Incomplete punches are another source of headaches for the user of the machine. Keypunches do not malfunction to the point where depressing the A key produces the code for Z, but incomplete punches often occur when duplicating. If the letter A is being duplicated and only the 1 portion of the code is correctly determined, then a 1 will be punched. Machines that malfunction in this way continually need to be serviced. Use another punch, or manually repunch the card.

When two keys are pressed simultaneously, this will occasionally result in a bad punch. These can be avoided by "clean" finger strokes. Correct the error by duplicating and inserting a card in the usual way. Note that some of the more modern machines have a keyboard rollover feature. Up to four simultaneous punches can be remembered, which allows the machine to catch up with the operator.

A rare but frustrating problem occurs when the backspace key jams. It does not lock, but it can catch on the keyboard housing. Try a gentle pull upward to remedy the situation.

One last problem is misalignment during duplication. The card to be duplicated must be inserted squarely or else it will jam or cause reading problems. A code that is read incorrectly will be punched incorrectly. Cards inserted with the column-80 edge first or in another nonstandard way will, of course, also result in incorrect duplication.

Our discussion of common keypunch problems ends here. However, two important issues remain. Although they are not technically problems, an informed keypunch user will be aware of both areas.

First, let us take a brief look at the discrepancies between the codes and symbols used by keypunches and those used by printers. Printers may be equipped with printing elements (print chains, etc.) that produce any conceivable symbol in response to a specific code. Thus, the 12-1 code for the letter A might result in an infinity symbol or even a Japanese character. There are usually no problems with letters and numbers. It is the special characters that cause almost all of the problems. For example, printers are often equipped with special characters appropriate to accounting operations. If you receive output of this nature, there is no cause for alarm. It does not indicate that you have made an error in keypunching but rather that the print chain is not standard. There is no danger of misinterpretation. A 12-1 code is always the same. The symbols on either end of the input-output process do not matter. It is the code that counts.

This situation is most often encountered in computer centers where printing chores are handled by an older and less costly computer. In this situation, there is usually a provision specifying that the printing be done on a specific device. Only in unusual circumstances will these printer discrepancies be a problem. Your friendly computer consultant should be able to help you.

A short examination of the ecological impact of the keypunch is the final topic. Keypunches consume significant amounts of electricity. The typical keypunch room is very noisy, and it is easy to forget that a machine is still on. There can be no question that turning off idle keypunches is sound and responsible behavior. Card waste is also something that many students do not think twice about. Even experienced users ruin hundreds of cards in a year. Most computer centers have special bins for ruined cards. Recycled cards add up, so do not throw your mistakes out. Some computer centers also have special bins for waste printer output. Use them! It is just common courtesy to clean up the area for the next user of the machine.

5.8   The Future

One need not have mystical powers to see into the future of keypunching. Punched cards are doomed to fade away and be replaced by other, more direct input systems. For example, it is now possible to key directly onto magnetic tape. Cards are eliminated entirely. Such systems use a televisionlike CRT screen and a special keyboard. The user inputs a tape instead of cards. At the moment, high initial set-up and maintenance costs prevent wide use of these systems. They are economically feasible only where volume is very high and speed is crucial. The terminal input system is rapidly replacing the others. Even if you currently use a card system, you would still be wise to read Chapter 4 on terminals because there is no question that terminals will one day be used nearly everywhere.

Part Two begins with a chapter devoted exclusively to input, the single most troublesome area in computer-aided analysis. Anyone new to this field should be certain to understand the principles of input in this chapter before going on to learn about the analysis packages. This is true even if you will be using data that have been provided for you, since familiarity with these principles will contribute to a firm understanding of data sets and their preparation.

Chapters 7, 8, and 9 are devoted to the three most commonly used statistical packages: SPSS, SAS, and BMDP. These packages overlap by 40–60%. The chapters are similarly structured in order to illustrate the many things that these packages have in common. Ideally, one should learn all three packages, although few researchers are familiar with more than two. If your goal is to add another package to your current repertoire, then you can probably skip Chapter 6. Go directly to the chapter you need.

Chapters 7 and 9 (SPSS and BMDP) are not truly complete without Chapter 6 (or a sound understanding of basic input procedures). Chapter 8 (SAS), on the other hand, uses input that is sufficiently different from that for the other two to warrant a separate discussion (see Section 8.2). This is one of the reasons that, in general, SPSS and BMDP are the two packages most commonly used.

PART TWO

Each of the chapters contains an applied section, which is an opportunity for practice. The manuals, while they do provide examples, have two inherent problems. They are generally incomplete, providing little sense or insight into the process itself. Also, the research problems are not familiar to many students.

It has been suggested that the problem of transfer among research domains is a trivial one—that the blood chemistry example used throughout BMDP can represent most social science areas simply by changing the names and labels. Perhaps this is true, but the changing of labels is an extra burden on the student/researcher, and confusion may arise in complex applications. Since this transferability is a highly controversial point, here are some specific instances. First, certain analysis packages define the term variable much more liberally than do many social scientists. Thus, the number of variables in a study is different from the number in the computer analysis. Second, the means of hypothesis testing vary widely among disciplines, and this results in some unintelligible examples. Third, some disciplines such as political science are oriented to data bases, and others such as laboratory psychology are oriented to single (or, at best, series) experiments. The experiment-oriented disciplines have a vital interest in files; the data-based professions have no interest in files at all.

The solution to the muddle of definitions and goals is to provide a wider selection of examples. Accordingly, in Chapters 7, 8, and 9, there are many exercises, which alternate between the survey and the experimental orientation. Each exercise treats only a few subjects or respondents. The decision to proceed in this manner was based on the assumption that if someone can make the program function correctly for 10 cases, then it can be done for 1000 cases. In effect, there are two sets of exercises, and the beginner must choose which are personally the most meaningful. Advanced users, of course, can see the bigger picture and probably will do some of each type. As was true for previous exercises, it is useful to attempt the solution using both this text and the relevant manual and then to check against the text solution. There is not a separate exercise for each program because many are very similar in operation and function.

Those readers who plan not only to work these applied exercises but also to actually run them as statistical analysis problems should use the expanded data sets contained in Appendix A. The data given in the exercises are far too brief to have much statistical meaning. All of the data are fictitious, and conclusions about real-world issues should be avoided.

84

Each chapter has a section called "Pro Tips and Techniques," which demonstrates advanced techniques and documents places of common confusion and problems. These suggestions and hints are not definitive; they are the products of experience. Many of these topics are also incorporated into the applied exercises. If a particular tip is not clearly understood, using it in an applied section might be helpful. The reader is encouraged to move back and forth between the applied sections and the tips and techniques section.

Before the discussion of the packages themselves, it should be noted that what follows is essentially a series of qualifications, exceptions, and complications. Part of the complexity of these packages stems from the improvements and additions that have been made to existing capabilities over time. Identical procedures are sometimes changed from version to version, or a procedure may be universal for all but one special program. Statistical analysis may not change, but the analysis packages do so constantly. When reading an analysis package manual, you will find that it is a good idea to write down the exceptions and qualifications as you go along. In anything as complex as most of these manuals, it does not help to make notes in the margins because you can never find them. An excellent place for notes is on the page at the end of the chapter in the manual or after one of the tips and techniques sections of this book. It is also helpful to index the manual with a set of looseleaf tab indexes.

6

Formats

After the control statements have been created, the computer must be instructed about the structural characteristics of your data, that is, the physical arrangement of the data (terminal lines, cards, tape, paper, forms, etc.). This is called the data format, or simply, the format. All programs and program packages contain data format provisions that range from the highly sophisticated DATA LIST specifications of the SPSS package to the long strings of numbers that reserve specific computer memory positions in machine language. Surprisingly, the more sophisticated a package is, the easier it is to use. The data input procedures particular to SAS, BMDP, and SPSS are of most use to the social scientist.

Most, if not all, analysis packages use or have the option of using the input specifications found in the FORTRAN or PL/1 programming language. In FORTRAN, the user would say to the machine, in effect, "Okay, machine, I want you to read from device number x the following variables (data) according to format number y." Standard FORTRAN recognizes three basic types of information: numeric (numbers only), noninformation (blanks and skipped columns), and alphanumeric (text). These are, respectively, the I and F formats, the X format, and the A (or literal) format. A specific analysis package may allow only certain of these formatting options. There are more FORTRAN formatting options, but discussion of these is left until the tips and techniques section of each program discussion.

Some computer-analysis specialists advocate complete elimination of formatting. In its place, they suggest the exclusive use of free formats with terminal data entry. Free formats require only that each datum be separated by a blank; beyond this, the data are formless. The author does not subscribe to this view, and so a more structured approach to data entry will be presented here. The reasoning behind this has two bases. First, when it comes to the physical arrangement of data, terminals and cards are the same. Terminals as well as keypunches require that data be described with perfect accuracy. The advantages of terminals are primarily those of ease and speed of preparation. Second, free-format procedures are not appropriate for the beginner. Such procedures are very prone to errors (both of omission and commission), and it is very difficult to proofread for these errors. These arguments are discussed in greater detail in this and in succeeding chapters. Free formats are fine for the experienced operator who knows how to allow for shortcomings and to recover from errors.

The examples and discussions in this chapter deal with both cards and terminal lines. With formatted data entry, the location of each datum on the 80-column line (either card or terminal line) is crucial. The physical differences have little importance; a datum in column 13 is described in the same way in card and terminal input. Each computer record is reproduced here to make clear the exact location of each character and each blank.

6.1   User Sophistication

There are beginning and advanced levels of formats in user-written programs. In packages such as BMDP and SPSS, only the input provisions of FORTRAN are used. Those readers who are familiar with the options and rules for FORTRAN *output* should totally forget these rules or at least should be careful not to use them as operational guidelines for statistical analysis.

In this text and in other introductory manuals, more space is devoted to formatting than to any other single topic. There are four major reasons for this:

1. Knowledge of other aspects of analysis packages is meaningless without the ability to properly input information.
2. The details of input vary among packages and even among specific programs within a package. Thus, a grasp of formatting procedures is required.
3. The error-checking routines in FORTRAN and in PL/1 compilers are able to detect only certain, very specific, problems. For ex-

ample, a given routine might specify that there are a hundred pieces of information on an 80-column card.

4. It is perfectly possible to input information that can be success-fully analyzed but is meaningless. A researcher desiring to obtain a correlation between intelligence quotients and geographical locations might enter the IQ information as a single digit—7 in-stead of 117. The program package would make the calculations and print the results as if no error had been made.

The user of analysis packages must have a thorough understanding of input procedures. If what goes in is meaningless, then what comes out will be no less meaningless.

6.2 Input Vocabulary

As discussed previously, the language used by computer specialists is unique and is not normally part of the social science researcher's vo-cabulary. This section provides the basic vocabulary needed to input (submit) a typical job to a computer using one of the major analysis packages.

record: A terminal line or a card line. While theoretically a record can be of any length, a length of 80 columns is fairly standard. *Note:* When less than 80 columns are transmitted, most systems fill out the remaining columns with blanks. In this text, the term *line* (as in data line) is used as a synonym for data record.

formatting: The process of informing the computer about the or-ganizational pattern or structural characteristics of the data to be input.

input format: The specific instructions about data, which are sub-mitted to the computer. (See Sections 6.3 through 6.5.)

output formatting: The process of specifying the structural prop-erties of the information generated by the computer analysis. Usually printed, output format may also consist of punched cards, terminal lines, tape, a disk, or all of these.

format card: The card or the terminal lines containing the input and output instructions.

format specifications: A series of conventions in FORTRAN or in PL/1 that denote the type of information to be input or output. For example, format specifications might state that data are in the form of whole numbers, integers, or numbers with fractional parts, or alphabetical information.

delimiters: Symbols (punctuation marks) used to note structural properties (other than the data themselves) of formats, such as the beginning and the end.

error messages: Information (generated either by the FORTRAN compiler or by the program package) that indicates an error in input, internal manipulations, or output. Such messages are often cryptic and difficult to interpret, but they are certainly better than no warning at all.

variable format: The FORTRAN term for the technique in which the format written by the package user is input to the program itself.

variable format card: Same as format card. The word variable refers to the fact that different arrangements may be used for different data; i.e., the arrangements are not fixed. When terminals are being discussed, the comparable term is *variable format instruction.*

field width: The number of places any given number occupies. The number 7851 occupies four places and has a field width of four. The number −.401 occupies five places and has a field width of five. Minus signs and included (not implicit) decimals count as digits when determining field width.

6.3   Types of Formats

Input procedures as they are used in individual packages will be reviewed in Chapters 7, 8, and 9. The number of exceptions and the overall complexity of these procedures necessitate a general introduction. Each discussion of an input format is followed by a section of applied exercises. It is important for novices to work through these exercises. The experts probably can skip the exercises and review only those areas that are of doubt. Intermediately-skilled researchers must make their own judgments about completing the applied exercises.

6.3.1   The X Format

The X format is the procedure for dealing with blanks, or spaces. Literally, it is a way to instruct the program to ignore certain portions of the input, to skip blanks, and to avoid considering blanks as information. This format has the general form of nX, where n is the number of characters to be skipped. Thus, 53X would cause 53 positions, or columns, to be ignored. Spaces are considered to be fixed at one unit in width. There is no 20X3 specification, for example. Skipping does not begin in column 17 for a 17X specification; rather, 17 columns are skipped. The format is evaluated column by column from left to right. If the first ten columns contain data and we give them a data format (see later sections) followed by 17X, then the skipping starts in column 11 and continues through column 27.

6.3.2 X Formats Applied

Data can rarely be described with a single format. The usual situation is a series of individual formats separated by commas. For the moment, we will delay the discussion of commas and other delimiters (the "grammatical" separators). Further, you should try to ignore the data and concentrate on the blanks. It will be very useful to you to write your formula in the margin or on scratch paper. Since the correct answer appears after the data, a ruler, sheet of paper, or index card should be used to cover the answer. In this way, you will be able to check your skill.

═══════ EXERCISE 1 ═══

Write an X format for a data line with the following pattern:

0123456789   0123456789

   Answer:

If we use the word data as a substitute for the numeric portion of the card (thereby allowing us to delay the discussion of numeric formats), we can write a format for the above record as follows:

```
data,5X,data
```


Patterns of data are not often so regular.

═══════ EXERCISE 2 ═══

Write an X format for the following data record:

012 345   02

   Answer:

A correct format is as follows:

```
data,1X,data,2X,data,69X
```

It is unnecessary to indicate in the X format when the remainder of a partially read record is skipped. The 69X in the format in Exercise 2 is superfluous; thus,

```
data,1X,data,2X,data
```

is also correct. The X format allows any part of a line, regardless of content, to be ignored.

───────── EXERCISE 3 ─────────────────────────────

Write a format for the following data assuming that you wish to ignore the first 18 numbers and begin with the 19th.

```
71264531961733A427659127434991
```

> *Answer:*

While any of the three following formats would be correct, the third one is preferable.

```
3X,15X,data
```

```
3X,15X,data,50X
```

```
18X,data
```

It does not matter what data you skip as long as the proper number of columns are ignored. The X format is not used when more than 79 columns are to be ignored. There is a separate method used for ignoring a complete record, which will be discussed later. In a two records-per-case situation, we do not skip to column 21 of the second record with a 110X specification. Instead, we use the special record-skip provision and a 20X specification.

Although X formats are straightforward, there are problems that commonly occur. Table 6.1 lists some guidelines that, if followed, will help you to avoid most mistakes.

6.3.3   The I Format

The I 'format is used for integer data, i.e., for numbers without fractional parts. This format has the general form nIw, where n is the number of repetitions of the same format and w is the number of digits required for each unit of data (w is also called the field width). IQ is a typical unit of data that consists of 3 digits. Thus, 25I3 tells the computer that there are 25 data units of 3 digits each and that they

TABLE 6.1. Guidelines for X Formats: nX

Dos	Don'ts
1. Carefully count columns to be skipped.	1. Do not use specifications larger than 79 columns.
2. Use 44X instead of 22X,22X.	2. Do not use field-width designators, such as 4X7.
3. Use X format when you want to ignore any portion of a record.	3. Do not use 0X or bX or X alone.
4. Remember that the number indicates the number of columns to be skipped *not* the number of the column to be skipped.	4. Do not use decimals, such as 7.0X.

are integer numbers. IQ is an excellent example of a typical social science integer; one never sees an IQ of 117.36. Numeric codes for gender are another example (0 and 1 or 1 and 2, typically.)

In general when we deal with social science data, there are almost no restrictions on the size or number of I formats; in other words, 4I20, 1I6, and 23I1 are all acceptable. Of course, the integer format and the actual data must match. The format 2I3 does not agree with the numbers 118.47 and 221.66.

6.3.4   I Formats Applied

In this section, you should try to write formats for each example and then check them against the given answers. Even though I formatting seems logical and probably error-free, it is not. Practice is strongly advised.

═══════ **EXERCISE 4** ═══════

Write an I format for a data record containing the number 951 in the first three columns.

951

   Answer:

A correct integer format would be as follows:

1I3

Clearly, 1I2 and 1I4 would specify incorrect field widths.

════════ **EXERCISE 5** ════════

Write an I format for the following data record, considering it to be a single number.

951274

    Answer:

The correct format would be this:

1I6

════════ **EXERCISE 6** ════════

If we think of the data in Exercise 5 as the two separate numbers 951 and 274, a different format is necessary. Write down what you think it is.

    Answer:

2I3

If we consider it as consisting of 6 one-digit numbers, then 6I1 would be the correct format.

════════ **EXERCISE 7** ════════

Consider a data record that has a three-digit number followed by a five-digit number followed by four one-digit numbers, as shown. Write a format instruction for this data record using commas as separators.

061254362130

    Answer:

1I3,1I5,4I1

Note that the four single-digit numbers are not related to each other in any particular way. It is correct but cumbersome to write an individual format for each. An individual-format instruction for the above example would be as follows:

1I3,1I5,1I1,1I1,1I1,1I1

Failure to correctly specify the field width of a number will result in a variety of errors. If 2I3 in Exercise 6 were 2I2, for example, instead of reading two numbers three-digits wide, the computer would read two numbers two-digits wide. The data used would be 95 and 12 rather than

951 and 274. Any result drawn from such a format would be totally erroneous in that the number 12 extracted from the middle of the sequence 951274 has no meaning. Of course, lack of meaning does not prevent the computer from calculating statistical differences, and the unsuspecting researcher might well conclude that the first set of numbers is different from the second set despite the fact that the second set is composed of nonsense numbers.

Errors in I formats are not necessarily readily detectable. If you are using *t*-tests, for example, the error will not be apparent in the test or in the test means. Only a comparison of the raw data with the input data will show the error.

================ EXERCISE 8 ================

Write a format for the following data. (Hint: The decimal place counts in determining the field width.)

651.7

Answer:

If you wrote 1I5, 1I4, or 1I5.1, you overlooked the fact that 651.7 is not an integer. It is not possible to write a correct I format containing a decimal. The format type must match the data. The error would be detected as a format error and would cause the computer to abandon your job, often with an irritating message such as "Bye, bye." (See the section on error messages in the chapter devoted to the particular analysis manual you are using.)

An acceptable (but not a recommended) variation of the I format is ϕIw. The number of repetitions is considered to be one—not zero. Social scientists routinely cause themselves much grief by misusing the 1Iw format. It is good procedure to insert the leading 1 whenever it is implied rather than to come to grief over such a simple matter. Once formatting becomes a matter of routine, discontinue the leading 1 as a matter of convenience, speed, and space-saving in the format instruction.

================ EXERCISE 9 ================

Write a format for the following data:

-67149

Answer:

If you wrote 1I6, you are correct. The format 1X,1I5 would work, except the number would be considered positive. The minus sign must be included in the determination of field width. Positive numbers do not need the plus sign, however; they will function correctly without it.

═══════ **EXERCISE 10** ═══════

Consider the case of multiple negative numbers. Write a format for these data:

871-541-61-890-771

Answer:

The correct specification is this:

1I3,1I4,1I3,2I4

Table 6.2 lists some guidelines for I formats. Reviewing this table is a good way to consolidate your knowledge of this format.

TABLE 6.2. Guidelines for I Formats: nIw

Dos	Don'ts
1. Use composites, such as 3I2 instead of 1I2,1I2,1I2.	1. Do not use numbers with any decimal places; 67.0 is not an integer.
2. Restrict I formats to integers.	
3. Correctly match the specified column width to the data width.	2. Do not use specifications larger than 80 columns, such as 152I3.
4. Count all minus signs when determining the field width; also count any plus signs needed.	3. Do not use 0Iw, nI0, 0I0, ᵇIw, nIᵇ, ᵇIᵇ, or I alone.
	4. Do not use decimals in the multiplier: for example, 3.7I4.

Note: There is an upper limit to the size of an integer that can be handled by a given computer. However, this limit varies among computers. On IBM 360/370 computers, the number is exactly 2147483647 *(IBM System 360/370 FORTRAN IV Language,* IBM Corp., 1974, p. 15).

6.3.5   The F Format

The F format is the most widely used specification. It denotes decimal numbers, including those having zero decimals. It has the general form *n*F*w.d,* where *n* is the number of repetitions, *w* is the field width for a single data unit, and *d* is the number of decimal places required. A person's height and a research animal's weight are typical examples of F-format data. The field width required is determined by the accuracy of measurement. Thus, if we were to record the heights of the Los Angeles Lakers basketball team to the nearest tenth of an inch, we would need a 14F3.1 specification. Translation? There are 14 real numbers (numbers with fractional properties) that are three-digits wide with the last digit representing the decimal portion.

Many of the principles that apply to I formats also apply to F formats; for example, the specification cannot refer to more than 80 columns, and decimal multipliers are not allowed. However, the F format is unique in that, despite the fact that this format is intended for use with decimal numbers, a decimal need not be entered into the data. This is what we mean by an implied decimal, and it is the most common way to enter data. In short, it is a waste of time to enter decimals if the computer can be told where the decimal place is located by some other means. Thus, if the data are 8429 and the F format is 1F4.2, the computer will automatically know that there are two decimal places, starting with the right-most digit. The computer understands the number to be 84.29 even though the decimal does not appear in the data.

If the decimal is entered on the data record, it must be counted into the field width. If the data card contains the number 4.9753, then the field width is 6, not 5. The same procedure applies to negative numbers. The minus sign must be counted in the field width. For example, −4.9753 has a field width of 7. There is a critical difference between these procedures; entering the decimal is optional, while entering the minus sign is mandatory. If the minus sign is left out, the number is read as positive. As discussed in Chapters 4 and 5, minus signs must be included, and, generally, decimals need not be.

The best way to learn F formatting is through practice, and Section 6.3.6 is recommended for those who have little experience. Readers requiring a shorter review of F formats should go to Table 6.3.

6.3.6   F Formats Applied

——— EXERCISE 11 ———

Consider a data record containing the number 24 followed by the number 7.5. Write an F format of two parts, using the comma to separate them.

247.5

98 COMPUTER USAGE FOR SOCIAL SCIENTISTS

Answer:

A correct format would be

1F2.0,1F3.1

If we consider the 24 to be 2.4 without the decimal, the format would be

1F2.1,1F3.1

Consider a situation wherein the data in Exercise 11 were read with the incorrect format 1F2.1,1F2.1. The effects of incorrect field size depend upon what precedes and what follows the data in question. In this case, the second piece of data will be read as 7., and a zero decimal will be assumed, i.e., 7.0. If the 24 is the age of a subject and if the 7.5 is the number of pounds lost in a weight-reduction experiment, the half-pound difference could be crucial, depending on the remainder of the data. Errors of this type can be critical.

=== EXERCISE 12 ===

Suppose that immediately following the weight data of Exercise 11 is the annual income of the subject ($12,300). Write a format for this record:

247.512300

Answer:

The correct format is as follows:

1F2.0,1F3.1,1F5.0

The following format is incorrect:

1F2.0,1F2.1,1F5.0

A good way to check your skill is to determine the effects of the incorrect format on the data given in Exercise 12. Give yourself a small pat on the back if you thought that the age would be read correctly and that the weight would be read as 7.0. If you thought the income (12,300) would be read correctly, you do not pass go and you do not collect $200! You should have realized that since the weight in the incorrect format is only two digits wide, the first digit of the income would be the 5, making the income read as 51,230–a very considerable difference!

The effects of incorrect formats are just as harmful when the specification is too large rather than too small. Consider the following format:

1F2.0,1F4.1,1F5.0

What will be the weight and the annual income if the data in Exercise 12 are read with this format? The weight becomes 7.51, and the income becomes 2500ƀ. The difference between 7.5 and 7.51 is slight and is therefore unlikely to affect most social science results. However, since a blank is read as a zero, the income becomes 25,000–again a considerable difference.

Occasionally, the problem is not with the format itself but with the data. What is wrong with the following data–format combination?

724856 13,500 17.50

2F3.0,1X,1F6.0,1X,1F5.2

If at first it appears that there is nothing wrong with this combination, it is because the problem is subtle (but very important). *Hint:* Look at the data. Never enter commas into the data![1] F formats accept numbers, plus and minus signs, decimals, and nothing else. A comma is a special character and is never part of a number. This comma problem is unusual, but a mismatch between repeated field widths is common.

================== EXERCISE 13 ==================

Consider the following data records to be 8.21, 7.35, 6.37, 7.01, 9.51, 3.99, and 5.55 without the decimals. Write a format for this series using the multiplier feature.

821735637701951399555

Answer:

Both of the following formats work, but the second one is inefficient.

7F3.2

1F3.2,1F3.2,1F3.2,1F3.2,1F3.2,1F3.2,1F3.2

---

[1]Note that there are valid exceptions to this rule but that they are not recommended for beginners. See Chapters 7, 8, and 9.

A series of numbers must be completely regular to correctly use the multiplier feature. Deviations will create problems as serious as field-size problems. The multiplier feature can be used to skip blank spaces if they occur in a regular fashion.

════════ EXERCISE 14 ══

Write both a conventional and a multiplier format for the following data:

821 735 637 701 951 399 555

     Answers:

7F4.3

and

1F3.2,1X,1F3.2,1X,1F3.2,1X,1F3.2,1X,1F3.2,1X,1F3.2,1X,1F3.2

Note that in the second answer to Exercise 14 the final 1X is omitted. While it is unnecessary, including it is acceptable. Readers with experience in FORTRAN format language will see another acceptable multiplier format for this example. Discussion of this option will be delayed until we examine the use of parentheses in formatting.

════════ EXERCISE 15 ══

As a final exercise in F formats, write a format for the following data (there is one of everything): 7.30, 7.21, .0334, −77, −6.4, −629. 761, +.7, 29, 43, 78, 56, −4.7, −8.2. Here is the data record:

730 721 0334 −77 −6.4 −629.761 +.7 29437856−47−82

     Answer:

In the F format that follows, some of the blanks are added to the numbers that precede or follow them. Experience demonstrates that sooner or later this practice will cause trouble and is used here only to make the examples complete.

2F4.3,1F4.4,1X,1F3.0,1X,1F4.1,2X,1F8.3,1X,1F3.1,1X,4F2.0,2F3.1

If you wrote your format statement in Exercise 15 using I formats for the −77, 29, 43, 78, and 56, you deserve congratulations. The es-

TABLE 6.3. Guidelines for F Formats: nFw.d

Dos	Don'ts
1. Use F formats for both implied and actual decimal data.	1. Do not use specifications such as 91F2.1 that are larger than 80 columns.
2. Use F formats for integer data as if it had an implied zero.	2. Do not use 0Fw.d, nF0.d, bFw.d, nFw.b, or F alone.
3. Use composites, such as 4F2.1 instead of 1F2.1,1F2.1,1F2.1,1F2.1.	3. Do not use decimals such as 4.5F3.0 in the multiplier.
4. Correctly match the specified column width. Minus signs, plus signs, and decimals count in determining field width.	4. Do not use I formats for decimal data.

Note: Exercise caution when using extrawide formats to cover blanks in the data.

sence of formatting is correctly matching data and format. Indeed, most formats are mixtures of I, F, X, and our next topic—the A format. Review Table 6.3 for F format guidelines before moving on to A formats.

The following list gives some special FORTRAN format terminology:

1. *Real numbers* are those that can have decimal values; for example, 4.2 and 8.0. The decimal need not be included on the card.
2. *Integers* are whole numbers, that is, numbers lacking fractional parts. For instance, 7, 528, and 929174 are integers.
3. *Floating point* for our purposes refers to decimal, or real, numbers. In FORTRAN, it refers to the type of arithmetic that is performed, as in "floating (decimal) point arithmetic."
4. *Fixed point* refers to arithmetic done with integers only. The distinction is made because integer arithmetic is faster.

Thus, we can see that F formats are floating point formats and that I formats are integer formats.

6.3.7   The A Format

Another commonly used format is the alphanumeric, or A, format, which has the form *nAw*. As with previous formats, *n* is the number of repetitions of the field width *w*. This format is often used for textual material, e.g., the name of the subject, a city, or mixed letters and numbers. Arithmetic is not done on A-format information; the mean of a list of city names makes no sense.

The field width of an A format is also limited in that it must be greater than zero and equal to or less than four. Be careful here, as the maximum width varies among computers and individual computer centers. Ask your consultant. The choice of a field width is somewhat arbitrary for the social scientist. All of the following A formats are correct ways to input HELENA, the name of the capital of Montana: 1A4,1A2 or 3A2 or 6A1. Each of these formats stores information differently because storage is based on internal functions. On IBM equipment, storage is normally in units of four characters. When A1 is specified, the remaining characters are filled with blanks. Table 6.4 illustrates this point.

TABLE 6.4. *The Effect of the A Format on the Storage of Data*

These data	*with this A format*	*result in data stored in this form in:*			
		Unit 1	*Unit 2*	*Unit 3*	*Unit 4*
HELENA	1A4,1A2	HELE	NA♭♭		
HELENA	3A2	HE♭♭	LE♭♭	NA♭♭	
HELENA	6A1	H♭♭♭	E♭♭♭	L♭♭♭	(etc.)

However, since A-format data are never analyzed, the method of storing data is not of great importance to the social science user. Remember that 3A2 and 6A1 denote 3 and 6 separate variables, respectively, to the computer. And 1A4,1A2 denotes two variables for an analysis package—not an English proper name. Thus, when A-format information is used to classify data, all the variables that make up the name must be used. For the data in Table 6.4 by the format 3A2, classification would be by HE, LE, and NA. (*Note:* Not all packages allow names as classifications.)

Any combination of letters, numbers, and special characters may be used. For example, HELENA1 and HELENA2 might be used, since there is also a Helena, Arkansas. Use of the special characters found in SCL or in other system instructions is acceptable but not advisable.

A formats are so straightforward that an applied section including exercises is not necessary. Practice with this format can be gained using the material in Sections 6.4 and 6.5. Table 6.5 lists the general guidelines for A formats.

TABLE 6.5. Guidelines for A Formats: nAw

Dos	Don'ts
1. Count decimals, signs, numbers, and letters when determining field width.	1. Do not use decimals in the multiplier or in the field width.
2. Avoid SCL symbols in the data, especially the slash and the asterisk.	2. Do not use specifications greater than 80 columns, such as 40A3.
3. Remember that each multiple counts as a separate variable.	3. Do not exceed the maximum field width available (in most circumstances, it is four).
4. Remember that A-format variables are never used in arithmetic operations.	
5. Remember that A-format variables can be used as classification variables in some but not all packages.	

6.4   Format Delimiters

The formats discussed thus far are adequate for most social science input requirements as long as those formats can be used in conjunction with one another. As we all know, social science data are rarely uniform in structure or in organization. Data in uneven pieces can be input to an analysis package by using delimiters. These are special characters that begin (the left parenthesis) and end (the right parenthesis) a whole format or that separate individual formats (the comma), for example, (5X,2F3.1,1X,4F1.0).

So far, this chapter has broadly treated the topic of formatting. However, each analysis package has its own formatting requirements. For example, BMDP requires that the above format be written:

```
FORMAT = '(5X,2F3.1,1X,4F1.0)'.
```

SPSS requires the following format:

```
INPUT FORMAT FIXED (5X,2F3.1,1X,4F1.0)
```

Between the parentheses the formats are the same and will convey the same information to the computer.

In the following sections, we begin with a discussion of parentheses and commas. Then we will deal with using a slash (/) as a delimiter. Experience indicates that interpreting slashes causes problems for many researchers, so a special section will be devoted to that subject.

6.4.1   Commas and Parentheses

Commas separate individual specifications. Thus, if the age and the sex of a subject are in the first three columns of the data, two separate formats are required. The formats 1F2.0 and 1F1.0 or two I formats could be used. In either case, a comma is used to separate the two formats, as follows: 1F2.0,1F1.0. Commas separate all formats regardless of their type.

Commas are not used to begin or end a format. This is the function of the left and right parentheses, as in this example:

(5X,4F3.1,4A4,10I1,1X,1I2.1)

Parentheses and commas are both delimiters and are not used consecutively. Therefore, you should not use ,) or),1. Parentheses may *not* be substituted internally for commas, as in this example:

(5X(4F3.1(10I1(1X(1I2)

The one exception to the internal use of parentheses is their use in delimiting a format that is to be multiplied several times, e.g., (5(1F2.0)). Note that the format begins and ends with parentheses in addition to those for the internal multiplied format. The number of left parentheses must always equal the number of right parentheses. This process is referred to as parentheses compounding, or embedding. Parentheses may be used in multiples and in compound compounds:

(5X,12(1F1.0,1X)10(2F2.0,1X))

or

(5(2(F2.0,1X)))

Note that in these examples commas and parentheses do not appear adjacent to each other.

One topic remains to be discussed before moving on to the applied uses of commas and parentheses. The compounding of parentheses must not produce a specification greater than 80 columns; i.e., not more than one complete record. The slash is the multiple-record delimiter. For a complete discussion of the slash, see Section 6.4.3. It is quite simple to check for the 80-column limit. Add all of the formats together (considering all multiples), and get the grand total. Table 6.6 gives some examples that demonstrate this process.

TABLE 6.6. *Effective Record Lengths of Compound Formats*

Format	Column and Multiplier	Total Columns
(5(2F2.0,1X))	$5 \times (2 \times 2 + 1)$	25
(16X,2(2I4)6(5F1.0,2X))	$16 + 2 \times (2 \times 4) + 6 \times (5 \times 1 + 2)$	74
(2((2A4,1X)(3I2,1X,4F2.1)))	$2 \times ((2 \times 4 + 1) + ((3 \times 2) + 1 + (4 \times 2)))$	
	or	
	$2 \times (9 + 15)$	48
(16X,4F2.1,1X,2F6.2,1X,10A2)	$16 + (4 \times 2) + 1 + (2 \times 6) + 1 + (10 \times 2)$	54

You should also always make sure that the number of left parentheses in a format equals the number of right parentheses. In the formats in Table 6.6, there are two pairs in the first example, three in the second, four in the third, and one in the fourth.

6.4.2   Commas and Parentheses Applied

As mentioned earlier, practice is essential, especially now that you are ready to write a complete format. Table 6.7 lists some guidelines for using commas and parentheses. These rules may seem obvious, but they reflect the most common mistakes. More experienced users should review Table 6.7.

TABLE 6.7. Guidelines for Commas and Parentheses

Dos	*Don'ts*
Commas	
1. Separate individual formats with a comma.	1. Do not insert extra commas.
2. Make sure commas are commas, not periods.	2. Do not interchange decimals and commas.
3. Remember that blanks are separators in some format systems such as SAS but that commas are separators in all systems.	3. Do not separate multipliers with commas, as in (17X,4,(2F2.0)).
	4. Do not delete required commas.
	5. Do not substitute the apostrophe (') for the comma.
	6. Do not put commas next to parentheses. (The exceptions to this rule are dangerous for the inexperienced to attempt.)
Parentheses	
1. Remember to count the right and left parentheses; they must be equal.	1. Do not insert extra parentheses.
2. Remember that 3(F2.0), 3(1F2.0), and 3F2.0 are equivalent. Use the simpler form of 3F2.0.	2. Do not substitute the vertical bar (logical OR) for the parentheses.
	3. Do not substitute left for right parentheses, or vice versa.

━━━━━━ **EXERCISE 16** ━━━━━━━━━━━━━━━━━━━━━━━━━━━━━━━━━━━━━

Write an F format for the following data, considering the first group of digits to be a case ID number:

0100 178 180 180 179 177 175 174 173 173 171 170 170

Answers:

There are a variety of correct formats:

(A4,1X,12(1F3.0,1X))

(1A4,1X,11(1F3.0,1X)1F3.0)

(2A2,1X,12(1F3.0,1X))

(1A4,1X,1F3.0,1X,1F3.0,1X,1F3.0,1X,1F3.0,1X,1F3.0,1X,1F3.0,1X,1F3.0,1X,1F3.0,1X,
1F3.0,1X,1F3.0,1X,1F3.0,1X,1F3.0)

The first three formats are more efficient, but they are also more difficult to check against the actual data. The last format is easy to compare to an actual data record, and hence the danger of miscounting columns, and so forth, is lessened. The last format also demonstrates another important principle—that many formats are actually longer than a single record. Each program package allows continuation onto succeeding lines, but each does so in a slightly different way.

The formats in Exercise 16 could have been written with I formats. For this exercise, however, we use F formats, since they are more general and since some packages do not allow the I format at all.

━━━━━━ **EXERCISE 17** ━━━━━━━━━━━━━━━━━━━━━━━━━━━━━━━━━━━━━

In order to make sure that the entire concept of formats has become clear to you, write a format for the beer-drinking contest information in Table 6.8. The data are not real because the contest was not actually conducted. However, these data are typical of social science data. Assume that you are not planning to analyze the case numbers, dates, or names, but that you do want this information available for labeling purposes. In other words, you will use A formats for case numbers and dates. The top of the table lists the columns in which the data would be entered if they were real. Use these columns to determine your format.

Answer:

A correct format would be

(1A2,1F2.0,1F2.0,1F1.0,1X,1F2.1,1X,1A4,1A3,1X,1F2.1,1X,1A4,1A3,1X,1F2.1,1X,1A4,
1A3,1X,4A4)

TABLE 6.8. *Summary of Data for Beer-Drinking Contest at Kelly's Bar*

Case	Age	Class	Sex		L beer/30 min	Date		L beer/30 min	Date		L beer/30 min	Date		Name
00	00	0	0	1	01	1111111	1	22	2222222	2	33	3333334	4	4444444455555555555
12	34	56	7	8	90	2345678	9	01	3456789	01	12	4567890	1	2345678901234567
01	19	80	1		37	27SEP77		43	17JAN78		40	02MAY78		C Reilly
02	18	80	2	B	11	23SEP77	B	23	15JAN78	B	31	29MAY78	B	S O'Grady
03	61	78	1	L	41	26SEP77	L	42	14JAN78	L	40	05MAY78	L	S Williams
04	20	79	1	A	21	24SEP77	A	23	16JAN78	A	19	04MAY78	A	A Amato
05	20	78	2	N	30	23SEP77	N	39	16JAN78	N	42	05MAY78	N	D Palito
06	19	79	1	K	19	20SEP77	K	23	16JAN78	K	23	05MAY78	K	D Stein

(See Appendix A for more data.)

Note: L = liter, to the nearest tenth. 1 = male; 2 = female.

═══════ **EXERCISE 18** ═══════

If in Exercise 17 you didn't write a format that takes advantage of the repetitive nature of the Kelly's Bar data, do so now.

Answer:

Such a format would appear as follows:

(1A2,2(1F2.0)1F1.0,3(1X,1F2.1,1X,1A4,1A3)1X,4A4)

A minor variation would be

(1A2,2(1F2.0)1F1.0,1X,3(1F2.1,1X,1A4,1A3,1X)4A4)

Note that the test-date data are divided into two parts because of the limit on A-format size. The same is true of the names at the end of the table.

6.4.3   The Slash

The final delimiter is the slash (/), or the end-of-record indicator. It directs the computer to cease processing the current record and to go on to the next. The slash is simply the skip command. In this sense, it is similar to the X format. However, the slash is a delimiter, and it must not be set within commas. For example, the format (2X,4F6.1,/,7F2.1) will not function correctly; (2X,4F6.1/7F2.1) is correct. Similar to the parentheses, the slash may appear adjacent to other slashes or parentheses as follows:

(/2X,4F6.1,3(1F2.1,1X)/)

Slashes may appear at any points where they are needed.

Proper use of slashes is often difficult for the beginner. For this reason, we will go directly to an extensive series of examples. In each case, you should attempt to determine the operation of the format instructions and which (if any) records will be skipped. It is essential that you work through each example, since the effect of the slash varies with its position in the format.

6.4.4   The Slash Applied

The exercises in this section present simplified input data to allow you to concentrate on the slash. Also, the data are described rather than printed as they were in previous exercises. This should not be a problem with such simple data. Finally, remember that decimals are not entered.

Write a format instruction for data consisting of three two-digit variables with one decimal place, beginning in column 1. Instruct the computer to ignore cases 1, 3, 5, etc.

Answer:

(/3F2.1)

The above format instructs the system to input (read) the cases as follows:

1. Skip the first record and begin reading on the second record.
2. Read three numeric variables that are each two columns in length and have one digit to the right of the decimal.
3. Skip the third record, and begin reading on the fourth. . . .

The general effect is one of inputting only the even cases.

Write a format instruction for 11 variables, each four digits wide with two decimal places and beginning in column 1. Skip the first and second data records, read the third, etc. Analyze the effects of this format, as in the answer to Exercise 18.

Answer:

(//11F4.2)

This format instructs the system to:

1. Skip the first and second records and begin reading on the third record.
2. Read 11 numeric variables each four columns in length with two digits to the right of the decimal.
3. Skip the fourth and fifth records, begin reading on the sixth. . . .

The general effect is to read every third record.

Write a format for data containing three two-digit variables, each having one decimal place, followed by a second record of four two-digit variables having no decimal places. Variables begin in column 1 on both records. Analyze the effects of the format. Do not use the X format.

Answer:

(3F2.1/4F2.0)

This format instructs the system to:

1. Read three numeric variables, each two columns in length and having one digit to the right of the decimal from the first record.
2. Skip to the second record, and read four numeric variables each two columns in length and having no digits to the right of the decimal.
3. Read three numeric variables, each two columns in length and having one digit to the right of the decimal from the third record. . . .

The general effect is to read a pair of records for each case.

═══════ **EXERCISE 21** ═══════════════════════════════════

Write a format for data containing two three-digit variables with two decimal places. Skip the second record. Read one variable of two digits with two decimal places. Variables begin in column 1. Analyze the effects of the format. Do not use the X format.

Answer:

(2F3.2//1F2.2)

This format instructs the system to:

1. Read two numeric variables each three columns in length with two digits to the right of the decimal from the first record.
2. Skip the second record, and begin reading on the third record.
3. Read one numeric variable two columns in length with two digits to the right of the decimal.
4. Read two numeric . . . from the fourth record. . . .

The general effect is to read three records per case.

═══════ **EXERCISE 22** ═══════════════════════════════════

Write a format for data where the first record has one two-digit variable with no decimal places, the second record has three two-digit variables with no decimal places, and the third record has nine two-digit variables, each having one decimal place. All records begin in column 1. Analyze the effects of the format.

Answer:

(1F2.0/3F2.0/9F2.1)

This format instructs the system to:

1. Read one numeric variable two columns in length and having two digits to the right of the decimal.
2. Skip to the second record, and read three numeric variables that are each two columns in length and have no digits to the right of the decimal.
3. Skip to the third record, and read three numeric variables, each having two columns in length and one digit to the right of the decimal.
4. Read one numeric variable from record four. . . .

Again, the effect is to read three records per case.

Slashes appearing at the end of a specification have the same function as those appearing at the beginning. Complete records are skipped. Exercises 23 and 24 are logically similar to Exercises 18 and 19. However, you should work through them to further establish your expertise with slashes.

===== **EXERCISE 23** =====

Write a format for data containing three two-digit numbers, each having one decimal. Skip the second record, and read the third record, etc. Analyze the effects of the format.

Answer:

(3F2.1/)

This format instructs the system to:

1. Read three numeric variables, each two columns in length and having one digit to the right of the decimal.
2. Skip the second record, and begin reading on the third record.
3. Read three numeric . . . .

The effect here is to read data from odd-numbered records.

=============== **EXERCISE 24** ===============

Write a format for data containing seven two-digit variables with no decimal places. Skip the second and third records. Analyze the effects of the format. All variables begin in column 1.

> *Answer:*

(7F2.0//)

This format instructs the system to:

1. Read seven numeric variables, each having two columns in length and zero digits to the right of the decimal place.
2. Skip records two and three, begin reading on record four.
3. Read seven numeric . . . .

Thus, data are read from the first, fourth, seventh, etc., records.

As a final exercise in this series, consider formats requiring beginning, embedded, and ending slashes.

=============== **EXERCISE 25** ===============

Write a format for data. Skip the first record. Read three two-digit variables, each having one decimal place from the second record. Skip to the next record, and read four one-digit variables with no decimal places. Skip the next record. Analyze the effects of the format. All variables begin in column 1.

> *Answer:*

(/3F2.1/4F1.0/)

This format instructs the system to:

1. Skip the first record, and begin reading with the second record.
2. Read three numeric variables, each two columns in length and having one digit to the right of the decimal.
3. Skip to the third record, and read four numeric variables, each one column in length and with no places to the right of the decimal.
4. Skip the fourth record.
5. Skip the fifth record, and begin reading with the sixth record.
6. Read three numeric . . . .

Four records will be read or skipped for each case.

Thus, the general rule for slashes is that slashes at the beginning or at the end of formats skip entire records and that embedded slashes skip to the next record.

The F format is obviously not the only specification that may be used with the slash. X and A formats may be used, as in the following:

(15X,3F2.1,3A2/7X,2F7.1)

Slash problems can be minimized by keeping the differences among beginning, embedding, and ending clearly and persistently in mind. Forgetfulness aside, errors are most common when slashes are used in conjunction with compound parentheses.

Such difficulties can be illustrated with the beer consumption data of Table 6.8. Assume that the decision has been made to execute the beer analysis with the odd-numbered subjects (reserving the even-numbered subjects for cross-validation, perhaps), and that the class and date information are not to be included. Thus, a format that skips every other record is required:

(2X,1F2.0,2X,1F1.0,3(1X,1F2.1,8X)/)

If this format had been written as follows:

(2X,1F2.0,2X,1F1.0,3(1X,1F2.1,8X/))

all kinds of problems would result. Exercise 26 documents the details of the format for the beer consumption data without slashes. Exercise 27 documents the effects of slashes outside of compound parentheses. Exercise 28 documents the effects of slashes contained within compound parentheses. It is essential that you work through these exercises. In each one, try to determine the number of records read and the number skipped.

========= EXERCISE 26 =========

Determine the effect of the format given in the following answer. Remember that the intent is to analyze only the odd-numbered functions.

Answer:

(2X,1F2.0,2X,1F1.0,3(1X,1F2.1,8X))

This format instructs the system to:

1. Skip columns 1 and 2; begin reading in column 3.
2. Read one numeric variable two columns in length and with no decimal places.
3. Skip columns 5 and 6.

4. Read one numeric variable one column in length and with no decimal places.
5. Repeat the following three times per case:
   a. Skip one column (number 8 on first repeat).
   b. Read one numeric variable two columns in length and with one decimal place.
   c. Skip eight columns (numbers 11–18 on first repeat, numbers 30–40 on last repeat).
6. Repeat instructions 1–5 for each case (subject).

What is the total number of records read per case? One.

═══════ **EXERCISE 27** ═══════════════════════════════════

Determine the effect of the format given in the answer below. Will it read only the odd-numbered cases?

Answer:

(2X,1F2.0,2X,1F1.0,3(1X,1F2.1,8X)/)

With this format, the first five instructions to the system are the same as those in Exercise 26. Then the system is instructed to:

6. Skip the second record, and begin reading in the third record.
7. Repeat instructions 1–6 for each case.

How many records are read and skipped per case? Two. This format instructs that only the odd cases are read.

═══════ **EXERCISE 28** ═══════════════════════════════════

Note that the subsequent format with embedded slash results in a somewhat bizarre skipping of cases. Determine the effect of the embedded slash format given in the answer below.

Answer:

(2X,1F2.0,2X,1F1.0,3(1X,1F2.1,8X/))

With this format, the first four instructions to the system are the same as those in Exercise 26. Then the system is instructed to:

5. Repeat the following specification three times per case.
   a. Skip one record column (number 8 on first repeat).

    b. Read one numeric variable that is two columns in length and has one decimal place.

    c. Skip eight columns.

    d. Skip to the next record, and begin reading.

6. Repeat instructions 1–5 for each case.

Note: On the last repeat, the slash is an ending slash and is not embedded. Thus, the final action is to skip an entire record (the fourth record on the first cycle).

How many records are read and skipped per case? Four.

The marked differences between Exercises 26, 27, and 28 are critical to correct analysis. Exercise 26 calls for the information about subject (case) 1 to be located on the first record, for information about subject 2 to be on the second record, and so on. Subject 1 is on the first record, but subject 2 is on the third record, and so forth, in Exercise 27. Note that consideration is not given to subject ID numbers but only to the sequence specified by the format. The information about subject 1 according to Exercise 28 is not contained on the first record but rather on the first, second, and third records, skipping the fourth record. Subject 2 is dealt with on the fifth, sixth, and seventh records, skipping the eighth record. The formats displayed certainly have very different effects. Keep in mind, however, that these effects are valid only in the sense that they may be interpreted by the computer without error.

In the case of the beer-drinking study, only the format in Exercise 27 is valid in the sense of correctly fulfilling the researcher's intention of analyzing just the odd-numbered cases. (Remember that there is only one record per case in this study.) If we assume that there are 100 subjects in this study, the format in Exercise 26 will read 100 cases, including 50 even-numbered ones. The format in Exercise 27 will read 50 odd-numbered cases. The format in Exercise 28 will read only 25 cases —neither the first 25 nor the last 25, but rather 25 cases comprised of bits and pieces in groups of four from the data deck. This illustrates a very serious error.

It might be argued that most packages would not complete the analysis requested in Exercise 28. The number of cases specified and the number read would not agree, causing the program to halt (terminate) the processing and to generate an error message. This argument holds true if the number of cases is indeed specified in the program control instructions. However, many packages allow the number of subjects to remain unspecified; that is, the program reads until an end-of-file instruction (SCL) is encountered. In such circumstances the format in Exercise 28 would not generate an error message; there would not be even the slightest hint that the program had found anything wrong. The central issue is that the common end-of-data indicators (FINISH, /*, and END) offer no protection against format specifications that are technically correct but that direct an analysis of the wrong data.

There is one final caveat. In Exercise 28, once the program has entered the innermost parentheses and has skipped to the second record, the next unit of data will be in columns 2 and 3—not in columns 9 and 10 as on the first record. Hence, not only would data from subjects 2 and 3 be read as beer data from subject 1, but this incorrect data would not even indicate the amount of beer consumed by subjects 2 and 3. A close examination of the format in Exercise 28 and Table 6.8 reveals that what is taken as beer information is in fact the ages of subjects 2 and 3. Here we see error upon error.

6.5   Review and Practice

Only two remaining obstacles stand between the researcher and complete and confident use of formatting procedures. The first, how to format according to the specifications of each individual package, will be dealt with as a special section of the chapters devoted to BMDP, SPSS, and SAS. The second is practice, including practice with format problems as similar as possible to those situations actually encountered. The following exercises are designed to simulate the actual input typical of social science. It is highly recommended that each format be written before the answer is reviewed. There is no other way to be certain about your knowledge and skill. Advanced users should move directly to the analysis chapters.

In order to present the following exercises in the most efficient way, several simplifying steps have been taken. Variables are indicated graphically rather than verbally. To write a correct format, one does not need to know that the number 675 is the microgram amount of a new antipsychotic drug given to patients in an institution. One needs to know only that the number is three digits wide, that it has a specified number of decimal places, and where it is physically located on the record or records. Remember that there are several acceptable solutions to most of these exercises; just be sure that your answers are equivalent! For each exercise correct formats are listed. But do not worry if there are minor differences between your solutions and those given. For example, reading the group ID from the second, third, etc., record in a multirecord situation is desirable in some instances and undesirable in others.

═══════ **EXERCISE 29** ═══════════════════════════════════════

Write a format for the data described in Table 6.9. Each variable is listed along with the number of digits and the number of decimal places required in a left-to-right sequence. Here is a sample record:

001 THOMPSON 675 49134 67616210543217 56766273343788877225 456235645623452358966

> *Answer:*

(1F3.0,1X,2A4,1X,1F3.2,1F2.0,1F3.0,2F2.0,1F3.0,1F1.0,3F2.0,3X,18F1.0,1X,1F1.0,
F2.1,29F1.0)

═══════ **EXERCISE 30** ═══════════════════════════════════════

Write a format for the data described in Table 6.10. Each variable is listed along with the number of digits and decimal places required in a left-to-right sequence. After completing the first format, write another format that will read variables 15-30 and 45-50. Here is a sample of the records:

0001 1 1 1 2 27 2 5 4 1 2 5 3 4 1 2 5 4 1 4 2 5 3 5 3 2 1 3 2 4 5 1 3 4 2 5 4 1

0001 1 1 2 2 5 4 1 3 5 2 1 4 2 1 5 3 5 5 4 2 1 4 3 5

> *Answers:*

(1F4.0,1X,4(1F1.0,1X)1F2.0,1X,32(1X,1F1.0)/5X,3(1F1.0,1X)21(1X,1F1.0))

and

(1F4.0,1X,4(1F1.0,1X)1F2.0,17X,16(1X,1F1.0)/5X,3(1F1.0,1X)8X,6(1X,1F1.0))

To avoid reading redundant information, skip the case ID on the second record. However, sometimes just such redundant information does save the day when cards are dropped, shuffled, or lost. On a terminal, card misplacement is not a problem, but there are ways to electronically scramble data. It is still good procedure to check for matching data categories.

═══════ **EXERCISE 31** ═══════════════════════════════════════

Write a format for the data described in Table 6.11. After completing the format for all variables, write a format for variables 10-19 and 80-95. The experimenter's initials at the end of each record should be skipped, and the ID should be read only once. Here is a sample of the records:

001 1 1 1   5 7 8 3 9 4 2 1 3 6 5 8 7 9 6 5 4 2 8 9 5 3 4 5 4 3 6 2 5 4 3 8 2 DMK

001 1 2   9 6 5 4 3 8 7 1 1 2 9 4 2 5 7 3 6 5 2 4 1 8 6 7 2 3 6 5 4 8 7 3 5 8 DMK

001 1 3   5 2 3 4 6 2 7 5 4 2 1 9 8 3 4 5 6 7 8 2 4 6 3 5 2 4   0 GRADY   DMK

TABLE 6.9

Variable	Number of digits	Number of decimal places	Function
1	3	0	ID
2	8	0	name
3	3	2	GPA (grade point average)
4	2	0	state code
5	3	0	local area code
6 & 7	2	0 each	age of father/mother
8	3	0	college major
9	1	0	gender of student
10–12	2	0 each	authoritarianism score
13–22	1	0 each	conservatism/liberalism score
23–31	1	0 each	conservatism/liberalism score
32	1	0	school class
33	2	1	interviewer ID
34–42	1	0 each	conservatism/liberalism score

TABLE 6.10

Variable	Number of digits	Number of decimal places	Function
First record			
1	4	0	case ID
2	1	0	group ID
3	1	0	group ID
4	1	0	record number
5	1	0	gender
6	2	0	age
7–38	1	0	questionnaire responses
Second record			
39	4	0	case ID
40	1	0	group ID
41	1	0	group ID
42	1	0	record number
43–63	1	0	questionnaire responses

COMPUTER USAGE FOR SOCIAL SCIENTISTS

TABLE 6.11

Variable	Number of digits	Number of decimal places	Function
1	3	0	subject ID
2	1	0	group ID
3	1	0	record number
4	1	0	gender
5-37	1	1	reaction times
38	1	0	group ID
39	1	0	record number
40-73	1	1	reaction times
74	1	0	group ID
75	1	0	record number
76-101	1	1	reaction times
102	14	0	subject names

Answers:

(1F3.0,1X,3(1F1.0,1X)1X,33(1X,1F1.1)/4X,2(1F1.0,1X)34(1X,1F.0)/4X,2(1F1.0,1X)
26(1X,1F1.0)2X,7A2)

and

(1F3.0,1X,3(1F1.0,1X)12X,10(1X,1F1.1)//12X,16(1X,1F1.0)16X,7A2)

───────── **EXERCISE 32** ─────────

Write a format for the data described in Table 6.12. After completing the first format, write another format for variables 20-30, 70-80, and 110-130. Write a third format for variables 1-10, 42-52, 79-89, 108-118, and 145-155. (It may seem repetitious to write so many formats, but there is a lesson to be learned from each of these traps for the unwary!) Here is a sample of the records:

```
001 1 1 1  5 7 8 3 9 4 2 1 3 6 5 8 7 9 6 5 4 2 8 9 5 3 4 5 4 3 6 2 5 4 3 8 2 DMK

001 1 2  9 6 5 4 3 8 7 1 1 2 9 4 2 5 7 3 6 5 2 4 1 8 6 7 2 3 6 5 4 8 7 3 5 8 DMK

001 1 3  5 2 3 4 6 2 7 5 4 2 1 9 8 3 4 5 6 7 8 2 4 6 3 5 2 4                DMK

001 1 4  4 3 1 2 5 5 4 2 4 1 3 4 3 1 4 2 5 5 1 3 4 1 3 4 2 5 1 3 4 1 3 4 1 1 DMK

001 1 5  2 1 4 3 5 1 2 4 2 3 2 1 2 3 1 4                                    DMK
```

TABLE 6.12

Variable	Number of digits	Number of decimal places	Function
First record			
1	3	0	case ID
2	1	0	group ID
3	1	0	record number
4	1	0	gender
5-37	1	0	survey responses
Second record			
38	3	0	case ID
39	1	0	group ID
40	1	0	record number
41-74	1	0	survey responses
Third record			
75	3	0	case ID
76	1	0	group ID
77	1	0	record number
78-103	1	0	survey responses
Fourth record			
104	3	0	case ID
105	1	0	group ID
106	1	0	record number
107-140	1	0	survey responses
Fifth record			
141	3	0	case ID
142	1	0	group ID
143	1	0	card number
144-159	1	0	survey responses

Answers:

(F3.0,1X,3(1F1.0,1X)33(1X,1F1.0)/4X,2(1F1.0,1X)34(1X,1F1.0)/4X,2(1F1.0,1X)
26(1X,1F1.0)/4X,2(1F1.0,1X)34(1X,1F1.0)/4X,2(1F.0,1X)16(1X,1F1.0))

and

(1F3.0,1X,3(1F1.0,1X)30X,11(1X,1F1.0)/4X,2(1F1.0,1X)58X,4(1X,1F1.0)/1F3.0,
2(1F1.0,1X)3(1X,1F1.0)/16X,21(1X,1F1.0)/)

and

(1F3.0,1X,3(1F1.0,1X)6(1X,1F1.0)4(/10X,11(1F1.0,1X)))

In this chapter, we have presented the overall input procedure that is common to most analysis packages. While the length of the chapter may have seemed somewhat wearisome, formatting is one of the most difficult skills to master. At this point, you should be able to handle any analysis package. However, the unique aspects of each package are yet to be explored.

7

SPSS:
Versions 6, 7, 8, and 9

The Statistical Package for Social Scientists (SPSS) was conceived in 1965 in response to researchers' frustrations with the available assortment of incompatible and operationally different programs. Since then, it has enabled many thousands of researchers to perform otherwise impossible analyses. Many have learned to use SPSS with only the manual for guidance, but their skills have not been gained without difficulty. This chapter is intended to make the learning process easier, but it should never be considered as a substitute for the manual.

Beginners should start with Section 7.1, which provides a structural overview of SPSS. Intermediately skilled users who are seeking to refresh their skills should begin at the point where they feel their knowledge is the weakest. Section 7.2 is devoted to input, and 7.3 to applied practice. For the researcher who is familiar with basic SPSS procedures and feels no need for specific practice, Section 7.4 reviews techniques and tips in areas that typically cause problems. Section 7.5 reviews each of the major analysis procedures with an eye to warning about some common pitfalls.

A final point to keep in mind is that four versions of SPSS (6, 7, 8, and 9) are in common use, and variations of each exist for specific computers and computer installations. A few questions to your computer-center staff may save you considerable frustration.

7.1   Basic Structure and Function of SPSS

It may well be that the most intimidating aspect of SPSS for the beginner is its thoroughness. The mountain of detail is quite formidable.
Although the blind approach of just following directions will work,
to attempt it is to risk total confusion. The purpose of this section,
then, is to describe the structure and function of SPSS without the
details.

In the chapters devoted to the other analysis packages, there is little
need to discuss the structure of the program packages themselves.
Structure, however, is critical in SPSS, since this package cannot be
partitioned into manageable units. When you request SPSS, you get the
complete package, and in that sense every analysis can potentially use
all the capabilities of SPSS. To help the beginning researcher cope with
the totality of this complex package, this discussion uses an approach in
which the complex SPSS structure is reduced to three pathways. Each
pathway is further reduced to its basic decisions. Then the three pathways are reunited (as they are in SPSS) into a single decision point—the
point where the actual analysis is determined.

7.1.1   Basic Pathways in SPSS

The first level of decisions that arise when using SPSS is mapped out in
Figure 7.1. After the decision to use SPSS has been made, the system
control cards (SCL) are prepared. Explicit advice about the preparation
is impossible, for each computer installation is different and the SCL
must be modified according to the type of use.

As can be discerned in Figure 7.1, the first SPSS instruction is RUN
NAME, and this instruction is followed by a decision about the location
of the data. The data are either already filed in SPSS or are provided by
the user. When data are supplied, one of two methods must be chosen:
DATA LIST or VARIABLE LIST. These options constitute the three
separate pathways through the SPSS system. They can be briefly described as follows:

Pathway I: Data are provided by the user on either terminals, cards,
   tape, or disk and are described to SPSS by naming the variables
   and by providing a format specification. Card input is the traditional method; terminal input has largely replaced the use of cards,
   however.

Pathway II: Data are provided by the user on either terminals, cards,
   tape, or disk, and are described to SPSS with a single list. This list
   combines the functions of naming variables and of formatting.
   Cards are the traditional method here, too.

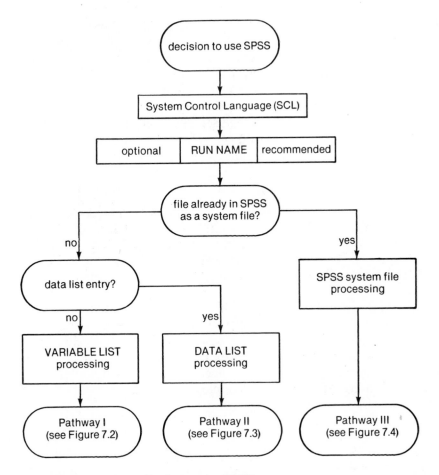

FIGURE 7.1. Basic Pathways in SPSS

Pathway III: Data are internally filed in SPSS and include both the names of the variables and their locations. Since the data are internal, the storage medium is generally of little interest. When an internal data file is processed, it remains intact and is changed only by special SPSS procedures.

However, experience indicates that the biggest problem for the beginner to avoid is getting bogged down in the myriad of choices. At the same time, the beginner must try to gain a thorough understanding of the large number of specifications required.

The details of input for each of these methods are taken up in the next section. Figures 7.2, 7.3, and 7.4 continue the pathway analogy. SPSS instructions are in capitals and are enclosed in rectangles. Whether an instruction is required, conditional, or optional is usually noted in the rectangle at the left. Additional information on the status of the instruction may also be included at the right. Ovals in each pathway are decision points that only you the user can describe. However, many are requests for options and should be avoided (i.e., choose "no") until you are familiar with the basic system.

7.1.2   Pathway I

Figure 7.2 illustrates the choices and requirements of Pathway I. Notice that there are three decision points that, if responded to with a "yes," lead to additional required instructions. In the instance of SUBFILE LIST, the number of cases is not specified. This is reasonable, since the SUBFILE LIST instruction requires inclusion of the number of cases per subfile.

In the SPSS system, each of these instructions is typed (or punched) one per line (or one per card) in a fixed location. These locations are called *control fields* and are always in columns 1 through 15. They must begin in column 1, and the spelling and spacing must be exact. Columns 16 through 80 are the *specification field,* which contains the details of each instruction. For example, in a study of 97 cases, we would have:

```
N OF CASES      97
```

Note that SPSS does not start or end instructions with special characters or punctuation marks. SPSS uses commas as logical separators in the specification field. As an example, in a study of the (supposed) relationship between college grade point average (GPA) and the physical characteristics of height, weight, eye color, gender, and age, the VARIABLE LIST instruction would be as follows:

```
VARIABLE LIST  ID,GPA,HEIGHT,WEIGHT,EYECOLOR,GENDER,AGE
```

Blanks also may be used as separators and may be mixed with commas. However, the names associated with each variable must not contain blanks or special characters and must begin with a letter. SPSS can distinguish between sequence names such as WEIGHT1, WEIGHT2, WEIGHT3, etc. Sequences may be implied rather than explicit, as in the following:

```
VARIABLE LIST  WEIGHT1 TO WEIGHT40
```

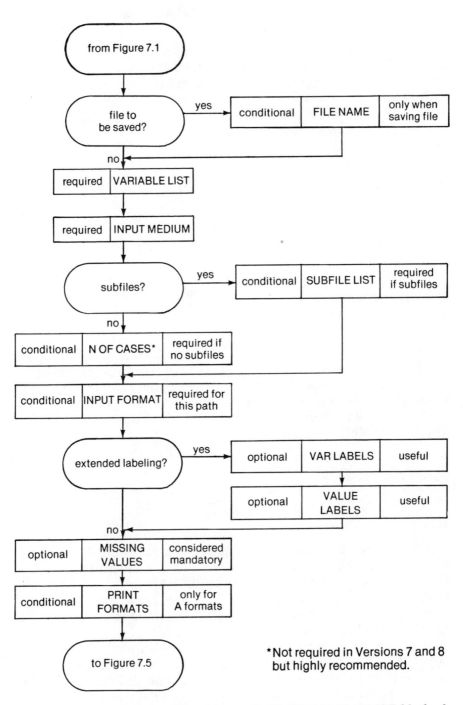

FIGURE 7.2. Pathway I: Files Not in SPSS, VARIABLE LIST Method

This is a considerable convenience in the case of long lists. However, the actual data must have the same arrangement (sequence) as the variable list implies. For example, if in the GPA and physical attributes study each subject (case) was weighed on the first day of class for four years and his or her GPA was recorded at the end of each year, the VARI-ABLE LIST instruction would be

```
VARIABLE LIST   ID,GPA1 TO GPA4,HEIGHT,WEIGHT1 TO WEIGHT4,EYECOLOR,GENDER,AGE
                HAIRCOLOR
```

There are clear limits to these implied lists. None of the following are acceptable:

```
GPAFR TO GPASR

1GPA TO 4GPA

GPAONE TO GPAFOUR

WEIGHT001 TO WEIGHT004

ITEM1 TO 15

GPA4 TO GPA1
```

The various instructions that make up Pathway I can be illustrated with the following hypothetical problem. Suppose that you are investigating the psychological profiles of Tolkien's hobbits Bilbo, Frodo, Sam, Pippin, and Merry with the *Minnesota Multiphasic Personality Inventory* (MMPI). The MMPI has 560 items.

```
RUN NAME        HOBBIT MMPI PROFILES
VARIABLE LIST   ID,NAME,ITEM1 TO ITEM560
INPUT MEDIUM    CARD
N OF CASES      5
INPUT FORMAT    FIXED(1A1,2A3,1X,72F1.0/5X,75F1.0/5X,75F1.0/5X,75F1.0/5X,75F1.0/
                5X,75F1.0/5X,75F1.0/5X,38F1.0)
MISSING VALUES  ITEM1 TO ITEM560 (0)
```

All of these instructions are required except MISSING VALUES, which is important enough to be considered mandatory. The input format should hold no surprises, for at this basic level it is merely the content of Chapter 6 in action. The simplest possible set of instructions is represented above. The discussion of options such as VALUE LABELS is included in the Pathway II subsection.

It is very important to take notice of a peculiarity in the INPUT MEDIUM instruction. The specification will be CARD whether the input is from keypunched cards or from a terminal. That is, SPSS will think you have submitted cards even if in fact what you have submitted are lines typed at a terminal. The above instruction and those that follow are appropriate both to terminals and to punched cards.

7.1.3   Pathway II

The principal difference between Pathways I and II lies in the use of the DATA LIST instruction instead of the VARIABLE LIST and INPUT FORMAT instructions, as noted in Figure 7.1. The data list is, in fact, a logical combination of the naming functions of the variable list and the data-structuring properties of the input format. Here are the instructions for the hobbit study using Pathway II in Figure 7.3:

```
RUN NAME          HOBBIT MMPI PROFILES
DATA LIST         FIXED (8)/1 ID 1 NAME 3-7 (A),ITEM1 TO ITEM72 9-80/2 ITEM73 TO
                  ITEM147 6-80 /3 ITEM148 TO ITEM222 6-80 /4 ITEM223 TO ITEM297
                  6-80 /5 ITEM298 TO ITEM372 6-80 /6 ITEM373 TO ITEM447 6-80 /7
                  ITEM448 TO ITEM522 6-80 /8 ITEM523 TO ITEM560 6-43
INPUT MEDIUM      CARD
N OF CASES        5
VAR LABELS        ID,IDENTIFICATION NUMBER/NAME,HOBBIT FIRST NAMES LAST NAMES
                  UNNECESSARY
VALUE LABELS      ITEM1 TO ITEM560 (1)YES (2)NO (3)UNABLE TO DECIDE
MISSING VALUES ITEM1 TO ITEM560 (0)
```

As previously mentioned, the details of SPSS input are taken up in Section 7.2. However, the data list above is readily decoded. FIXED specifies standard fixed-column input, and 8 indicates the number of records (terminal lines or cards). Then each variable on the first record is described in sequence, followed by the variable on the second record, etc. In this example, ID is the first variable and is in column 1, followed by NAME in columns 3 through 7. The (A) indicates that NAME is an alphabetic variable. The remainder of card 1 is filled with the first 72 test items (MMPI questions) in columns 9 through 80. Each succeeding record is then described in turn. Obviously, the VARIABLE LIST and INPUT FORMAT instructions are never used in conjunction with the DATA LIST instruction.

Several other features of SPSS are observable in the above example. Continuations always begin in column 16, and elements are never split. For example, a specification such as ITEM73 must not be divided between two records, but the string of specifications of which it is a part could be divided at any of the separators (delimiters), e.g., at the blank preceding ITEM73 or at the blank following it.

Also demonstrated here are the VAR LABELS and VALUE LABELS instructions. The first serves to decode the often (necessarily) cryptic variable names, and the second specifies the meaning of the values associated with each variable. The ID and NAME variables were not given value labels because it would be absurd to do so.

Pathways I and II are used with analyses of modest size in which the data are resubmitted each time. When there are a great many analyses or when the data set is large, it is very desirable to enter the data only

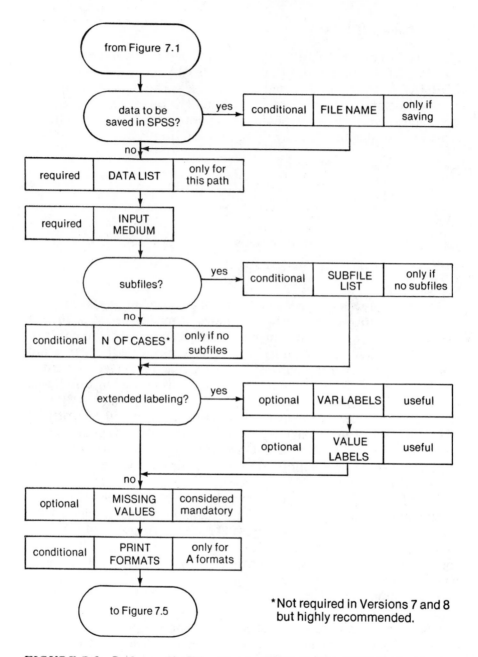

FIGURE 7.3. Pathway II: Files Not in SPSS, DATA LIST Method

once. In this case, either Pathway I or Pathway II is used for the initial entry. Subsequent analyses then use Pathway III.

7.1.4   Pathway III

Assuming that a file name was assigned in a previous analysis, following Pathway III is the essence of simplicity. One merely asks for the file, redefines variables if necessary, and moves on to specifying the tasks. The instructions outlined in Figure 7.4 would be as follows for the hobbit study:

```
RUN NAME        HOBBIT MMPI PROFILES
GET FILE        HOBBMMPI
MISSING VALUES ITEM1 TO ITEM560 (0,2,3)
```

The file name HOBBMMPI can be used here only if it has been assigned by a previous SAVE FILE instruction (see Figure 7.5). Note also that

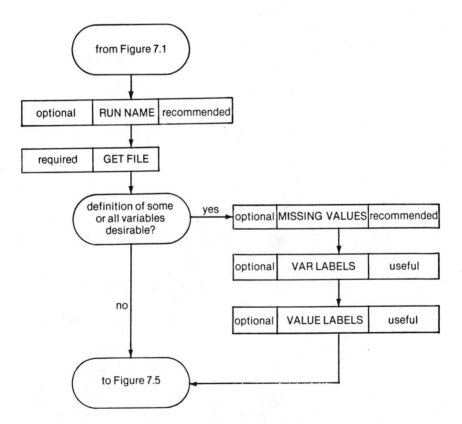

FIGURE 7.4. Pathway III: Files in SPSS, GET FILE Method

part or all of a file may be redefined. This does not change the original file unless additional instructions are stated.

There are two important points to remember about SPSS files. First, when files are used, special SCL instructions are required that of course are not part of SPSS proper. And second, some computer centers use modified versions of SPSS that do not allow files. For both problems, see your consultant.

7.1.5   Choosing the Analysis

The separate pathways merge when the actual analyses are to be specified. This combined pathway is outlined in Figure 7.5. The actual program selection is done with the TASK NAME instruction. The available programs are listed in Table 7.1 (see Section 7.1.6) and are referred to as tasks or subprograms. This last term leads to occasional confusion because there is no connection between subprograms and subfiles. Additionally, it is useful to refer to the subprograms as programs because it aids in the process of relating SPSS to other analysis packages. Structurally, SPSS is one large program.

Continuing the hobbit example, consider the following instructions for obtaining the frequencies associated with yes and no answers. The data are on terminal lines or cards.

```
TASK NAME        YES NO ITEM FREQUENCIES
FREQUENCIES      INTEGER=ITEM1 TO ITEM560(0,3)
OPTIONS          3
STATISTICS       1,5
READ INPUT DATA
(insert data here)
SAVE FILE        HOBBMMPI
FINISH
```

We can see from Figure 7.5 and from the above instructions that the SPSS user selects a task name and then a particular analysis. These choices are followed in turn by appropriately selected options and statistics. As many analyses as are desired may be produced by cycling through the sequence. Note that only one READ INPUT DATA instruction is required.

At this point, it should be clear that instructions such as RAW OUTPUT UNIT and RUN SUBFILES are used only where appropriate. It should also be evident that this sequence of cards is fixed. Frequent referral to the appropriate figure in this text will help you in the initial stages of mastering SPSS.

The analysis selection example above illustrates several new aspects of SPSS. The INTEGER instruction, as stated, will produce frequency tables for each item (up to 560 total items). This is not particularly

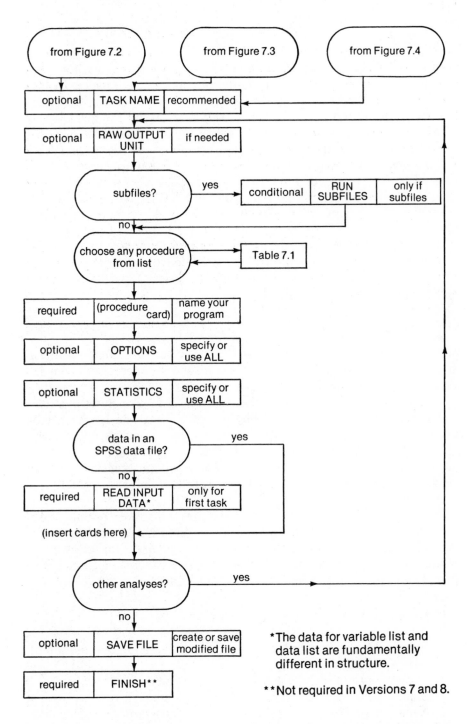

FIGURE 7.5. Task (Subprogram) Definitions for Pathways I, II, and III

133

useful in our example, since there are only five cases (five hobbits). It is all too easy to produce a mountain of meaningless paper. In this example, since the 3 in the OPTIONS instruction specifies that each table is to be printed in an $8\frac{1}{2}$-by-11-inch area (useful for inclusion in a manuscript), there would be 560 pages of tables. The statistics chosen above (mean and standard deviation) are also somewhat dubious. Recall that the answer yes was assigned a value of 1 and that no was assigned a value of 2. Yet there are only five hobbits! The beginning SPSS user should refrain from specifying ALL for the STATISTICS or OPTIONS instruction. The result in the hobbit example would be more than 1000 pages of nonsense.

Thus, it can be seen that the three pathways come together to form a single path with few complications, provided that one does not become involved in subfiles and in multiple analyses. The actual analyses available are the final topic of this section.

7.1.6 Available Analyses

Table 7.1 lists the major tasks that SPSS can perform. Many of these programs are complex analysis systems in and of themselves. For instance, the REGRESSION subprogram can execute a variety of regression analyses. All in all, there are thousands of unique analyses possible.

One question that is often asked about SPSS analyses is whether or not more than one analysis can be performed at once, such as CON-DESCRIPTIVE and SCATTERGRAM. The answer is yes, but the beginner should perfect his or her techniques before attempting simultaneous analyses. This is particularly true, since error detection and correction is an essential part of the learning process.

The examples given in preceding sections illustrate several points about SPSS that have yet to be mentioned directly. All analysis packages follow certain conventions for naming variables. The following is an abbreviated list of the rules for SPSS that was adapted from *SPSS*, second edition, page 30. (See your manual for more details.)

1. Length is limited to eight characters.
2. All names must begin with a letter.
3. Except for the first letter, the name may be any combination of letters and numbers.
4. Names may not include any special characters.
5. Blanks are considered to be special characters.
6. Names may not be split between a card (or a line) and its continuations.

TABLE 7.1. Available SPSS Tasks (Subprograms)

Functions available	Explanation
In all versions:	
1. CONDESCRIPTIVE	descriptive statistics
2. FREQUENCIES	frequency of occurrence stat
3. AGGREGATE	higher-order summations
4. CROSSTABS	two-way frequency tables
5. BREAKDOWN	crosstabulation with distributions
6. T-TEST	
7. PEARSON CORR	Pearson correlation
8. NONPAR CORR	nonparametric, or ordinal data, correlation
9. SCATTERGRAM	two-variable plots
10. PARTIAL CORR	partial correlation
11. REGRESSION	
12. ANOVA	analysis of variance
13. ONEWAY	one-way analysis of variance
14. DISCRIMINATE	multiple discriminate analysis
15. GUTTMAN SCALE	
16. FACTOR	
17. CANCORR	canonical correlation
In Versions 7 and 8 only:	
18. NPAR TESTS	nonparametric tests
19. MULT RESPONSE	analysis of multiple responses
20. RELIABILITY	tests theory reliability
In Version 8 only:	
21. REPORT	output utilities
22. SURVIVAL	life statistics
In Version 9 only:	
23. MANOVA	multivariate analysis of variance
24. BOX JENKINS	time-series analysis
25. NEW REGRESSION	multiple regression
26. GRAPHICS	statistical graphics option

Although it may seem obvious, spelling variations and abbreviations are not allowed. In effect, all of the instructions illustrated in this section must be used as given. Modifications will usually result in some form of error; see Section 7.5 for exceptions.

7.2   Input to SPSS

Input procedures can be divided into three categories: a VARIABLE LIST and a fixed format, a VARIABLE LIST and a free format, and a DATA LIST. The free format was not discussed in the preceding section, since it is an option and not a truly separate pathway. This section will present only a brief discussion of the fixed format because most of the requirements of a fixed format were discussed in Chapter 6. Free formats and the DATA LIST command are given a fuller treatment, as this is the first discussion of that combination.

7.2.1   The FIXED Format

The general form of the formatted input is as follows:

```
INPUT FORMAT   FIXED(any specification here)
```

The details within the parentheses are explained in Chapter 6. For example, we could have this:

```
INPUT FORMAT    FIXED(7X,14F2.1/25X,14F3.0)
```

or this:

```
INPUT FORMAT    FIXED(4A4,5(F1.0,1X)20F2.2)
```

The guidelines for correct formatting apply here. In addition, there are a number of unique requirements:

1. The INPUT FORMAT instruction must be in exactly this form and must begin in column 1. The word FIXED must begin in column 16.
2. I formats are not allowed.
3. The INPUT MEDIUM instruction must specify CARDS for both cards and terminals.
4. The location of these instructions within the PCL is fixed.
5. Multiple formats are not allowed, but continuations are acceptable. All continuations begin in column 16.
6. When continuations are necessary, a given specification may not be split between records.

Guidelines 5 and 6 are sufficiently troublesome to warrant an example. Consider the following specification:

```
INPUT FORMAT    FIXED(2F3.2,7X,F1.0,1X,3F3.0,5X,F7.4,1X,F8.2,1X,F4.3,1X,2F2.1,
                F3.1,6F1.0,2A2/7X,5F2.1)
```

If F3.1 were broken into F3. and 1, it would result in an error. (This is not true of other analysis packages.)

Of course, there must be agreement between the format and the data. Further, the order implied by the input format must correspond to that implied by the VARIABLE LIST instruction. Recalling the beer-drinking study from Chapter 6, we see that the following example demonstrates this necessary correspondence:

```
VARIABLE LIST   ID,AGE,SEX,LBEER1,DATE1,LBEER2,DATE2,LBEER3,DATE3
INPUT FORMAT    FIXED(1A2,2F2.0,1F1.0,1X,3(F2.1,1X,F6.0,1X))

0119831 37 800927 43 810117 40 810502
```

Deviations from exact correspondence do not necessarily produce error messages, but they certainly produce erroneous results. Generally, there is only one input format per SPSS run.

7.2.2   The FREE FIELD Format

Preparing a FREE FIELD format card is quite simple; only the instruction itself is required.

```
INPUT FORMAT    FREE FIELD
```

The sole restrictions are that the data must be separated by one or more blanks and that alphanumeric variables must be placed within apostrophes. For example, consider this data from the beer-drinking study:

```
01 19 83 1 37 800927 43 810117 40  'C REILEY
```

Blanks and apostrophes allow SPSS to identify variables from the variable list. Thus, as in fixed formats, the sequence of variable names is crucial. Data may also be separated by commas or by a mixture of commas and blanks.

```
01,19,83 1 37,800927,43,810117,40  C REILEY
```

Failure to insert commas or blanks will result in the interpretation of adjacent variables as one variable. Decimals must be punched, but there may be more than one case per record. For example, we might have this:

```
01 19 83 1 3.7 800927 4.3 810117, 4.0  'C REILEY       ' 02,20 82 1 3.9 800922 4.1
810201 4.1  'R DITOMMASO     ' 03 19 83 2 2.8 800915 3.6 810129 4.5  'T THOMPSON
' 04 (etc. . . .)
```

FREE FIELD formatting has several advantages for the SPSS user:

1. Detailed determination of field widths for each variable is not necessary.

2. Considerable preparation time is saved.
3. Rigid column structure need not be maintained.
4. No time is wasted waiting for a completed card to clear the keypunch.
5. Beginners can quickly learn to enter a small data set from a terminal.

There are, however, a number of disadvantages with FREE FIELD formatting:

1. Data prepared in FREE FIELD formats are generally much more difficult, if not impossible, to proofread.
2. Previously prepared data with adjacent fields (see preceding examples) cannot be read with FREE FIELD formatting. Of course those data could be read by one of the other procedures.
3. Alphabetic data not set in apostrophes are not readable with this procedure. Hence, alphanumeric data require extra effort and are still susceptible to error. Data previously prepared without apostrophes are totally unusable—even the alphabetic portions.
4. It is not possible to skip fields. Thus, the important advantage of being able to ignore certain utilitarian information is lost. And of course the apostrophe rule still applies.
5. It is not possible to skip cards in a multicard deck. It is also not possible to use the format specification to select only certain cases—all even-numbered cases, for example. This does not apply to tape, but it does apply to terminal lines.
6. A stray character in a data line will generally cause the entire job to be terminated because of the nonagreement between the variable list and the number of FREE FIELD variables. Such an error need not occur in each line; once is sufficient! (See the discussion below.)
7. Incomplete and blank lines not removed from the prepared data are not readily detected as errors. These lines are treated as more data by the FREE FIELD procedure. The same is true of blank and incomplete cards.
8. Decimals must be present; thus, a data set prepared initially without decimals cannot be read in a FREE FIELD format.
9. The use of format features associated with types I, X, P, E, T, D, G, and H is not permitted.
10. Error correction is much more difficult. Each data line (or card) must be individually scanned to find the place where the correction is needed. (See the "Pro Tips and Techniques" section for a solution to this problem.)

Proofreading difficulties are perhaps the most frequently encountered problem with SPSS FREE FIELD formatting.

===== EXERCISE 1 =====

Consider the raw data and card lines in Table 7.2 for ten cases from the morphine addiction and stress study in Appendix G. Proofread the lines (cards) in FREE FIELD format against the raw data. Doing so, you will see that it is quite difficult to maintain your place in the progression of numbers and that it is difficult to detect field-width errors visually.

TABLE 7.2. Raw Data and Lines (Cards) for Ten Cases

Variables	1	2	3	4	5	6	7	8	9	10
Case 1	3	031	044	2.01	2.45	076	2.29	2.37	1	03
Case 2	3	072	065	1.94	0.51	038	2.03	0.61	1	33
Case 3	4	011	024	1.75	4.12	076	1.70	4.04	1	15
Case 4	4	013	305	1.22	0.38	297	1.95	0.35	1	56
Case 5	4	291	304	1.50	0.51	306	1.91	0.48	1	43
Case 6	4	131	293	0.95	5.09	275	1.59	4.61	1	20
Case 7	4	022	015	0.87	3.47	037	1.41	3.99	1	25
Case 8	4	081	124	0.94	2.34	116	1.52	2.89	1	26
Case 9	4	132	135	1.37	0.21	147	2.00	0.43	1	31
Case 10	4	081	103	1.54	4.01	135	2.31	3.87	1	01

3 031 044 2.01 2.45 076 2.29 2.37 1 03 3 072 065 1.94 0.51 038 2.03 0.61 1 33 4
011 024 1.75 4.12 076 1.70 4,04 1 15 4 013 305 1.22 0.38 297 1.95 0.35 1 56 4
291 304 1.50 0.51 306 1.91 0.48 1 43 4 131 293 0.95 5.09 275 1.59 41.61 1 20 4
022 015 0.87 3.47 037 1.41 3.99 1 25 4 081 124 094 234 116 1.52 2.89 1 26 4 132
135 1.37 0.21 147 2.00 0.43 1 31 4 081 103 1.54 4:01 135 2.31 3.87 1 01

Answer:

In the lines (cards) above, there are six errors that you should have detected:

1. In the second line, the number 4.04 appears as 4,04. The effect of this error is to create two separate numbers. (Recall that both commas and blanks may be used as variable separators.) Thus, variable 8 receives a value of 4, variable 9 receives a value of 04, and variable 10 receives a value of 1. This error is then continued through all subsequent variables; i.e., variable 1 in case 4 has a value of 15.

2. In the third line, a value is listed as 41.61 when it should be 4.61. This error would probably go unnoticed with FREE FIELD formatting.
3. In the fourth line, we have 094 instead of 0.94.
4. Also in the fourth line, we have 234 instead of 2.34.
5. At the end of the fourth line, we have 4ɣɣ132 instead of 4ɣ132. The effect of this is that 132 will be merged with 135 on the next line. Again, the entire subsequent sequence of variables is disturbed.
6. In the fifth line, we have 4:01 instead of 4.01. This error would cause an error message to be printed, for a colon is a special character and is therefore detectable.

FREE FIELD formatting can be recommended only if caution is employed when cards are involved. When terminals are used, this format is faster. It is especially efficient with data for which the researcher is aware of the likely magnitude of results. The researcher can then use all the file-checking and screening abilities of SPSS before trusting the results. FREE FIELD formatting is deceptive; exercise caution.

7.2.3   The DATA LIST Option

As researchers' experiences with the early versions of SPSS accumulated, it became apparent that many user errors were caused by the difficulties of writing correct VARIABLE LIST and INPUT FORMAT instructions when input data were long and complex. The DATA LIST option is an attempt to provide an input procedure that is less prone to error. The following is a list of critical basic rules:

1. DATA LIST is not recommended as the first and only input procedure. The beginner should learn the VARIABLE LIST–INPUT FORMAT combination first.
2. DATA LIST input conforms to all SPSS conventions of variable naming, field widths, decimals, variable types, multiple records, and lists.
3. The efficiency and error-reduction properties of DATA LIST vary with each specific situation.
4. DATA LIST works well with multiple cards per case, but the default assumption is one record per case.
5. Records must be read in sequential order, but variables within a record may be read in any order.
6. The default assumption is that decimals, if any, are included with the data and that implied decimals and alphanumeric information will be specified.

The control field contains the words DATA LIST, and the specification field contains either the word FIXED or the word BINARY, followed by the specifications. FIXED is the standard mode, whereas BINARY indicates a data set written in internal machine notation. BINARY is a professional notation and is best ignored by everyone else. Immediately after the word FIXED is the number of records (lines or cards) per case. Thus, the following lines illustrate situations where there are 1, 1, 1, 4, and 13 records per case, respectively:

```
DATA LIST     FIXED/

DATA LIST     FIXED /

DATA LIST     FIXED(1)/

DATA LIST     FIXED(4)/

DATA LIST     FIXED(13)/
```

After the number of records has been specified, the variables from the first record of concern and their locations must be indicated. For example, if the variable MORPH was on the second of four records in columns 20 and 21, the DATA LIST card would be

```
DATA LIST     FIXED(4)/2 MORPH 20-21
```

One of the peculiarities of the DATA LIST option is the assumption that all necessary decimals have been included. Implied decimals (those not included) must be further noted in the specification. If we consider MORPH to have two decimal places, the instruction would be

```
DATA LIST     FIXED(4)/2 MORPH 20-23 (2)
```

If we have another variable HT (height), which is measured to the nearest tenth of an inch (for example, 58.2 inches) in columns 50–53, on record 4, the DATA LIST instruction would be

```
DATA LIST     FIXED(4)/2 MORPH 20-23 (2)/4 HT 50-53 (1)
```

A similar convention is used for alphanumeric variables; an A is inserted in place of the decimal indicator. Suppose that the variable NAME is on record 2 and in columns 1–19 in the preceding example. We would have:

```
DATA LIST     FIXED(4)/2 MORPH 20-23 (2) NAME 1-19(A)/4 HT 50-53 (1)
```

or:

```
DATA LIST     FIXED(4)/2 NAME 1-19(A) MORPH 20-23 (2)/4 HT 50-53 (1)
```

but not:

```
DATA LIST    FIXED(4)/4 HT 50-53 (1)/2 NAME 1-19(A) MORPH 20-23 (2)
```

Variables within a given record may be read in any order—presumably one that facilitates analysis. But once SPSS is directed to a record later in the sequence, the earlier records have been permanently skipped for the current analysis.

The features and applications of DATA LIST are best illustrated by examples. Consider three examples from Appendixes A and G (Beer-Drinking Study at Kelly's Bar, Morphine-Stress Study, Anti-establishment Attitudes in Off-Year Elections). The beer-drinking data specification would be

```
DATA LIST    FIXED / ID 1-2 AGE 3-4 CLASS 5-6 GENDER 7 LBEER1 9-10 (1)
             DATE1 12-17 LBEER2 19-20 (1) DATE2 22-27 LBEER3 29-30 (1)
             DATE3 32-37 NAME 40-55(A)
```

The morphine-stress data specification would be

```
DATA LIST    FIXED(1)/1 SIBS 1 BDATE 2-4 TDATE1 6-8 WTT1 9-12 MLMORPH1 13-16
             TDATE2 18-20 WTT2 21-24 MLMORPH2 25-28 GROUP 30 ID 31-32
```

The attitude survey data specification would be

```
DATA LIST    FIXED(2)/1 ID 1-4 GENDER 6-8(A) AGE 10-11 EDUC 13-14 STATE 16-19
             (A) CITY 20-25(A) Q1 TO Q26 26-77/2 Q27 TO Q30 5-12 NBHOOD 14-27
             (A) RACE 29(A) INCOME 31
```

These examples illustrate the use of actual and implied decimals, multiple records, spacing, and implied lists. All are critical. If, in the beer-drinking data specification, the blank between the variable GENDER and the column designator 7 were missing, SPSS would read a variable named GENDER7 and then would attempt to read a column designator. Not finding one, SPSS would suspend processing and end the run.

It is equally critical that variables in an implied list (e.g., Q27 TO Q30) be in adjacent columns. This means that the data pattern is regular and that the width of each variable is identical. Thus, Q1 TO Q10 23-52 denotes ten variables that are each three columns wide, and Q50 TO Q54 7-36 denotes five variables that are each six columns wide. Decimal points count toward the column total. Integer, decimal, and alphanumeric data types may not be intermingled. The morphine-stress specification illustrates how leading blanks used for purposes of readability can be accommodated. Trailing blanks serve to increase the value of the data by 10, by 100, etc. When the implied list (XXX TO XXX) is not being used, blanks can be skipped by simply not referring to any column(s) containing blanks.

A final feature of the DATA LIST option is how it skips records in comparison to the VARIABLE LIST–INPUT FORMAT option. For example, to skip to the 24th of 25 records requires this:

```
INPUT FORMAT    FIXED(//////////////////////// . . .
```

versus this:

```
DATA LIST       FIXED(25)/24 . . .
```

The form and the length of the data will determine whether or not the DATA LIST feature of SPSS is advantageous over other input options. It is most useful, as noted in the SPSS manual (2nd edition, pp. 53–55), when:

1. There are many records per case and many records to be skipped in a particular analysis.
2. The data are completely regular and of a uniform data type, such as all numeric and two columns wide.
3. Codebooks are available for the data, and their very length makes the automatic pairing of the variable and its record-and-column location valuable. (It is easy to get lost in format statements that are many records in length.)

The DATA LIST option should not be used when:

1. The repeat options of format statements lead to greater efficiency. For example, this format:

   ```
   (8F10.4/8F10.4/8F10.4 . . . )
   ```

   is more efficient than this one:

   ```
   FIXED (3) /1 MPOP1 TO MPOP8 1-80/2 MPOP9 TO MPOP16 1-80/3 MPOP17
           TO MPOP24 1-80/ . . .
   ```

2. Alphanumeric, integer, and decimal information is intermixed, particularly when the pattern is repetitive, making the multiple-parentheses option of the INPUT FORMAT instruction very useful. Mixed data preclude implied lists in the DATA LIST.
3. The user wants a shortcut to SPSS input procedures. The DATA LIST instruction may be less susceptible to error, as the manuals suggest, but it does not expose the student to the lessons of data types and data arrangement, and their relationship to the data records themselves.

The DATA LIST option is peculiar in the sense that SPSS translates the DATA LIST statement into VARIABLE LIST and INPUT FORMAT statements and then translates these into the appropriate instructions. Thus, all output from a run utilizing DATA LIST is accompanied by an INPUT FORMAT listing (on the first page of the SPSS output). It is entirely possible for a run to be successfully processed with the wrong input specifications. If the SPSS user is unable to read a format statement, there will be no way to determine if SPSS read the data correctly or not. The DATA LIST option is very useful, but it should be used only when the INPUT FORMAT specification is understood. It is not required that one be a whiz at formats; rather, one must be able to at least decode formats successfully.

7.2.4 Tape and Disk Input

Examination of Pathways I and II reveals that if the input is not on cards or on terminal lines then the previously discussed methods must be modified. When the data are not in an SPSS system file (Pathway III), input from terminals, cards, tape, and disk are nearly equivalent. Some form of variable naming and data formatting is necessary for all four means of input. The only required change is in the INPUT MEDIUM instruction. In effect, it is as if the cards were on tape or disk and therefore need to be described to SPSS. Combinations are also possible; for example, PCL instructions may be on terminal lines or on cards, and data may be on tape or on disk. In either situation, a READ INPUT DATA card is needed. In short, it does not matter where the data are in a physical sense; the procedures for Pathways I and II still apply.

Considerable flexibility is gained in this way. However, it does force the user to prepare system control language (SCL) that instructs the local operating system (not SPSS) about the characteristics of tape or disk files. The SPSS manuals provide examples of SCL for various installations, although these examples are not likely to be specific enough for actual use. Each local system varies to the point where a visit to your consultant is recommended.

When the Pathway III approach is used, VARIABLE LIST and other formatting instructions are not necessary because each SPSS system file already contains a complete description of the data. SCL instructions must be prepared, and of course there is no READ INPUT DATA instruction.

At first impression, it would seem that the use of SPSS system files is a great advantage because of the time saved by avoiding the long and detailed instructions characteristic of Pathways I and II. This is not so. A deck of cards is as much a permanent file as is an SPSS system file, and system files must be initially created the "long way." The real advantage of an SPSS system file is the speed of tape or disk over cards, especially when many analyses are contemplated. Many users will of course create data files on terminals instead of on cards. In this situation, the advantages of SPSS system files are limited. An evaluation of the relative advantages is beyond the scope of this text. Every option has its costs and its problems, and a careful examination of these issues will resolve the question for each combination of data, user, and operating system.

Some researchers have suggested that using Pathway III exposes them to fewer errors. This is true if someone else has prepared the system file for the user. It is not true when users prepare their own files. A correct data description on a terminal system is identical to a correct description on cards, tape, or disk. Again, the real advantages of input by Pathway III are speed and convenience.

7.3   SPSS Applied

The basic plan of the applied, or practice, sections was discussed at some length in the introduction to Part Two. Recall that each subsection is divided into two exercises; the first is for sociology and political science students, and the second for psychology and biology students. The two exercises always differ in substance but do not always differ in the SPSS issues raised. For example, treatments of MISSING VALUES will always be somewhat similar regardless of the discipline involved. Listed here are the exercises for SPSS and the specific topics introduced:

Exercise 2:   FREQUENCIES, INTEGER mode, implied lists, missing values, and Pathway I

Exercise 3:   FREQUENCIES, GENERAL mode, implied lists, missing values, and Pathway I

Exercise 4:   CROSSTABS, INTEGER mode, limiting output, Pathway I

Exercise 5:   T-TEST, correct comparisons, and Pathway I

Exercise 6:   SCATTERGRAMS, multiple subprograms, political science data modifications, and Pathway III

Exercise 7:   NPAR TESTS, multiple subprograms, psychology data modifications, and Pathway III

Exercise 8:   REGRESSION, COMPUTE, and RECODE instructions, and Pathway III

Exercise 9:   ANOVA, COMPUTE, and RECODE instructions, and
              Pathway III
Exercise 10: DATA LIST (Pathway II), complications and variations

All exercises should be completed for a thorough knowledge of
SPSS. This section is not intended as a review of all SPSS subprograms;
see Section 7.5 for information on specific programs. In all of these
exercises, it is essential that the order of SPSS instructions detailed in
Pathways I, II, and III be maintained. See Section 7.4.2 for further
details.

7.3.1   Practice with the FREQUENCIES Subprogram

You should begin learning SPSS with either the FREQUENCIES or the
CONDESCRIPTIVE subprogram. Both are descriptive in function; the
former subprogram applies to discrete data, and the latter applies to
continuous data. In many situations, data are composed of both types
and so require both subprograms. Recall that this does not necessitate
two separate runs—just two sets of task-definition instructions.
    The SPSS beginner should seriously consider the use of the EDIT fa-
cility. By specifying EDIT as the first instruction in columns 1 through
4, you can initiate the checking of your instruction syntax without
performing an analysis. Thus, you can be somewhat assured that things
are correct before an actual analysis is run. The EDIT facility does not
check for order or for illogical instructions.

═══════ EXERCISE 2 ═══════════════════════════════════════

Consider the following research question: Does the university or college experience
produce a political shift from conservative to liberal? Since the purpose here is to
practice with SPSS and not with research methods per se, the substantive qualifications of this question will be ignored. They are important but tangential.
    Each variable is listed in Table 7.3 along with the number of digits, the number
of decimal places, and the variable function. Assume this information is typed on a
terminal or punched on cards and that Pathway I has been chosen (VARIABLE
LIST and FORMAT). Six cases are listed; that is, the data are from six college students and their parents. Names and ID's refer to the student—not the parent. This
table is nearly identical to that in Exercise 12 in Chapter 6, so the format required
will be very similar. Read the description of the FREQUENCIES subprogram and
write the appropriate program control instructions using the INTEGER mode. Do
not analyze ID or NAME but do read them. We will take up how to save the data in
a system file in the next exercise.

TABLE 7.3. Practice Data on Shift from Conservatism to Liberalism

	Data Description		Function
Variable	Number of digits	Number of decimal places	
1	3	0	ID number
2	8	0	name
3	3	2	GPA (grade point average)
4	2	0	state code
5	3	0	locality code
6 & 7	2	0	age of father and of mother
8	3	0	college major
9	1	0	gender of student
10-12	2	0 each	authoritarian score for father, mother, and student
13	2	0	interviewer's ID number (father/mother)
14-22	1	0 each	nine conservatism/liberalism questions for father
23-31	1	0 each	nine conservatism/liberalism questions for mother
32	1	0 each	class in school (e.g., freshman)
33	2	0	age of student
34-42	1	0 each	nine conservatism/liberalism questions for student

Raw Data

```
001 SMITH      291 01107 58570341373614 01979868778 976678765 421375421142   DMK
002 PARSON     301 02004 65551092363830 06887889667 878877767 319742558677   DMK
003 VALENTI    389 01029 45430722282910 03774879989 766488995 420644311132   DMK
004 WALSH      400 04100 52590171252909 02657445753 445573667 219112134213   DMK
005 JOHNSON    210 01035 60590171394006 01897665998 887859767 320132234121   DMK
006 LARSON     237 01029 45470742292512 02655676687 998799887 118566458567   DMK
```

Note: These data begin in column 1 and end in column 80 when spaced as indicated.

Last of all, use the subprogram to obtain all of the descriptive statistics while you suppress the calculation and printing of frequency tables. The reasoning here is clear—one usually should obtain a description of all variables, but frequency tables need only be obtained for selected variables. After success with the descriptive statistics, try to obtain frequencies for the authoritarian scores, college major, and interviewers.

Answer:

```
EDIT
RUN NAME          POLITICAL ATTITUDE SHIFTS OF UNDERGRADUATES MAR81
VARIABLE LIST     ID,NAME1,NAME2,GPA,STATE,LOCALITY,AGEF,AGEM,MAJOR,SEX,AUTHF,
                  AUTHM,AUTHS,INTERID,FQ1 TO FQ9,MQ1 TO MQ9,CLASS,AGESTU,SQ1 TO SQ9
INPUT MEDIUM      CARD
N OF CASES        6
INPUT FORMAT      FIXED(1A3,1X,2A4,1X,F3.2,1X,F2.0,F3.0,1X,2F2.0,F3.0,F1.0,3F2.0,
                  1X,F2.0,9F1.0,1X,9F1.0,1X,F1.0,F2.0,9F1.0)
MISSING VALUES    GPA(9.99)/STATE(99)/LOCALITY(99)/AGE(99)/AGEM(99)/MAJOR(0)/
                  SEX(9)/AUTHF(99)/AUTHM(99)/AUTHS(99)/INTERID(9)/FQ1 TO FQ9(9)
                  /MQ1 TO MQ9(9)/CLASS(9)/AGESTU(99)/SQ1 TO SQ9(9)
VAR LABELS        AUTHF,AUTHORITARIANISM SCORE FATHER/AUTHM,AUTHORITARIANISM SCORE
                  MOTHER/AUTHS,AUTHORITARIANISM SCORE STUDENT
VALUE LABLES      STATE(1)PA,(2)NJ,(3)DE,(4)MD,(5)NY,(6)CN,(7)OH,(8)WV,(9)VA,(10)
                  MA,(11)DC,(12)UNKNOWN/SEX(1)MALE,(2)FEMALE/CLASS(1)FR,(2)SO,
                  (3)JR,(4)SR
FREQUENCIES       INTEGER=STATE TO AGEM AUTHF TO AUTHS AGESTU(01,99) FQ1 TO MQ9 SEX
                  CLASS SQ1 TO SQ9(1,9)
OPTIONS           7
STATISTICS        ALL
READ INPUT DATA
```

Beginning with the VARIABLE LIST instruction, we can see that it conforms to the various naming conventions and that implied lists can save considerable preparation. Writing FQ1 TO FQ9 is much more efficient than writing the nine individual names. Note that such lists apply only to adjacent variables. Alphabetic implied lists are not available in SPSS; i.e., one cannot write FQA TO FQI. Although these are legitimate variable names, the implied sequence is not recognizable. See below for a discussion of MISSING VALUES.

The next two instructions have no complications, and the only issue of note in the format is its length. Better planning of the data sequence would have resulted in a simpler format. But remember that the emphasis should be placed on logical and convenient data collection and not on convenient formats.

The MISSING VALUES instruction could be correctly written in several ways other than that given in this answer. Unlike the VARIABLE LIST command, this instruction allows nonadjacent variables having the same missing values to be grouped. For example, we might have this:

```
MISSING VALUES SAT(999)/STATE,LOCALITY,AGEF,AGEM(99)/MAJOR(0)/SEX(9)/AUTHF,
               AUTHM,AUTHS(99)/INTERID(9)/FQ1 TO FQ9,MQ1 TO MQ9(9)/CLASS(9)/
               AGESTU(99)/SQ1 TO SQ9(9)
```

Since the adjacency and the sequence of the data have been defined by the VARIABLE LIST instruction, we can group even further:

```
MISSING VALUES SAT(999)/STATE,LOCALITY,AGEF,AGEM,AUTHF,AUTHM,AUTHS,AGESTU(99)
               /SEX,INTERID,CLASS,FQ1 TO FQ9,MQ1 TO MQ9,SQ1 TO SQ9(9),MAJOR(0)
```

In effect, this specification states that there are only four types of missing values (999, 99, 9, and 0). With a very large survey, the time savings from such groupings are appreciable. When each variable has a unique code, none of these groupings is possible.

The VAR LABELS and VALUE LABELS instructions also may be grouped in the same fashion but only if the same labels apply to multiple variables. Indeed, a moment's reflection indicates that labels for some variables, such as labeling individual ages, would be silly. Unlike variable names, value labels may be up to 20 characters in length. Any legitimate characters except parentheses and the slash may be used.

Next is the task definition, or subprogram, instruction. The INTEGER mode was chosen over the GENERAL mode because it is significantly faster. However, the INTEGER mode is restricted to integers only. If it had been desirable to obtain descriptive information on ID's, then it would have been necessary to use the GENERAL mode because the A format is not allowed with INTEGER. The same is true for decimal numbers. Note that variables with the same maximums and minimums may be grouped as above.

In choosing OPTION 7, we eliminate all frequency calculations and tables. Thus, only those statistics requested in the next instruction are obtained. This affords a great deal of flexibility for the user of the FREQUENCIES subprogram.

It is important to note that all SPSS subprograms will function without either the OPTIONS or the STATISTICS instruction. Without these instructions, however, the default parameters are in effect, which may not produce the desired end. Also, recall earlier cautions about the use of the ALL option.

The FREQUENCIES subprogram has two very distinct modes; INTEGER is used in the preceding exercise, and GENERAL is used in the next. Since both modes apply to all phases of social science, it is highly recommended that you at least examine (if not actually work) Exercise 3.

═══════ EXERCISE 3 ═══

In this exercise, consider this question, which has perplexed educators for years: Is the test performance of certain students debilitated by high anxiety levels? Again, many detailed aspects of good research and design methods are omitted here in the interest of brevity.

The variables that were collected and their structural aspects are presented in Table 7.4. Assume that the information is on terminal lines or on cards and that Pathway I has been chosen (VARIABLE LIST and FORMAT). Eighteen cases are listed. These variables and cases are arranged in a two-factor study, with three levels

TABLE 7.4. Test-Anxiety Data

| | Data Description | | |
Variable	Number of digits	Number of decimal places	Function
1	3	0	ID
2	2	0	age
3	3	0	SAT math score
4	3	0	SAT verbal score
5	1	0	gender
6	1	0	class
7	3	0	major
8	3	2	grade point average
9	1	0	test-anxiety group
10	1	0	testing condition
11	2	0	final-exam grade
12	2	0	test-anxiety score

Raw Data

001	20	621510	14	031	319	1	1	55	21
002	19	490409	13	017	261	1	1	42	29
003	19	720800	12	019	385	1	2	89	25
004	22	750760	14	076	205	1	2	75	28
005	18	510498	13	085	322	1	3	69	30
006	18	390418	12	101	198	1	1	94	26
007	21	310370	14	064	285	2	1	82	18
008	20	650670	13	056	370	2	1	79	15
009	19	495390	13	034	360	2	2	98	19
010	19	620350	12	092	400	2	2	71	14
011	21	629510	14	091	265	2	3	96	16
012	18	530480	12	061	243	2	3	65	17
013	20	609610	13	020	301	3	1	85	09
014	21	720518	14	031	295	3	1	86	03
015	21	560540	13	041	261	3	2	79	04
016	19	490580	12	032	334	3	2	91	08
017	20	600590	14	005	325	3	3	82	09
018	20	510530	13	009	290	3	3	90	07

each. The first factor measures test anxiety and categorizes subjects into high, medium, and low test anxiety. Factor two consists of final-exam testing conditions: timed, unlimited time, and timed with "mood" music. Assume that the subjects were randomly assigned to the exam conditions. The dependent measure is the final exam grade.

As a first level of analysis, obtain complete descriptive statistics for all variables (excluding those that would be illogical) using the SPSS FREQUENCIES subprogram. After obtaining the descriptive information, obtain frequency tables for GPA, final exam grade, and test-anxiety score. Use the EDIT facility on your first try. (*Hint:* GPA is a variable having fractional properties.)

Answer:

```
EDIT
RUN NAME        TEST ANXIETY AND EXAMINATION CONDITIONS
VARIABLE LIST   ID,AGE,SATMATH,SATVERB,SEX,CLASS,MAJOR,GPA,TANXGRP,TESTCOND,
                FINAL,TANXS
INPUT MEDIUM    CARD
N OF CASES      18
INPUT FORMAT    FIXED(1A3,1X,F2.0,1X,2F3.0,2F1.0,1X,F3.0,1X,F3.2,2F1.0,1X,2F2.0)
MISSING VALUES  AGE(99)/SATMATH SATVERB(000,999)/SEX(9)/CLASS(9)/MAJOR(999)/
                GPA(999)/TANXGRP(9)/TESTCOND(9)/FINAL(00)/TANX(99)
VAR LABELS      GPA,GRADE POINT AVERAGE/TANXGRP,TEST ANXIETY GROUP/TESTCOND,
                TESTING CONDITION/FINAL,FINAL EXAM GRADE/TANXS,TEST ANXIETY SCORE
VALUE LABELS    SEX(1)MALE,(2)FEMALE/CLASS(1)FR,(2)SO,(3)JR,(4)SR/TANXGRP(1)HIGH
                TEST ANXIETY,(2)MODERATE TEST ANXIETY,(3)LOW TEST ANXIETY/
                TESTCOND(1)TIMED EXAM,(2)UNLIMITED TIME EXAM,(3)TIMED PLUS
                MOOD MUSIC
FREQUENCIES     GENERAL=AGE,SATMATH,SATVERB,SEX,GPA,FINAL,TANXS
OPTIONS         7
STATISTICS      ALL
READ INPUT DATA
(insert data here)
```

The VARIABLE LIST instruction here agrees with SPSS naming conventions, but is structurally different from the instruction in the previous exercise. Implied lists are useless because each variable is unique. See Exercise 2 for implied lists.

The INPUT MEDIUM and N OF CASES instructions need little explanation. A different arrangement of the data would have resulted in a simpler format, but remember that the emphasis should be on logical and convenient data collection—not on simple formats.

The MISSING VALUES instruction does not take advantage of the possibility of grouping variables with the same values. For instance, we could have

```
MISSING VALUES AGE TANXS(99)/SATMATH SATVERB(000,999)/SEX CLASS TANXGRP TESTCOND
               (9)/MAJOR GPA (999)/FINAL(00)
```

The instruction SATMATH SATVERB(000,999) could have been written SATMATH TO SATVERB(000,999), since these were defined as adjacent variables. This is of no real advantage here, but it would be if there were tens of hundreds of adjacent variables with the same missing values. Note also that several missing values may be defined per variable and that they need not be 9's or 0's.

Similar grouping may be done with VALUE LABELS and VAR LABELS instructions if the labels are to be repeated for a sequence of variables. Labels are not required. For some variables, they would be meaningless. Value labels may be up to 20 characters in length. Any legitimate characters except parentheses or the slash may be used.

In the next instruction—task definition, or subprogram selection—the GENERAL mode was chosen because data are comprised of a mixture of integer and continuous variables. Even alphabetic information can be accommodated by the GENERAL mode (A formats). However, this version of the FREQUENCIES subprogram runs considerably slower than does the INTEGER version.

As in Exercise 2, option 7 suppresses the calculation and the printing of frequency tables, producing only those calculations requested in the STATISTICS instruction.

The combination of the OPTIONS and the STATISTICS instructions provides a high degree of flexibility that allows the product to be tailored to specific needs. A most useful option is number 3, which causes all output to be printed in an $8\frac{1}{2}$-by-11-inch form for use in a manuscript. An OPTIONS instruction is not mandatory, but if you exclude it, be sure that the defaults are the ones you desire.

7.3.2   Practice with the CROSSTABS and T-TEST Subprograms

Social scientists at all levels often become embroiled in trying to answer substantive questions from a set of data before they have learned to use the tools of computer-aided statistical analysis. Resist the urge to get the answers immediately, and concentrate on the two topics introduced here: crosstabulations and SPSS system files.

=========== EXERCISE 4 ===========

Write the instructions for a crosstabulation between the genders of the students and their ratings on the conservatism/liberalism dimension. Use the data and variables from Exercise 2. Assume that we are using a self-rating that is the last of the nine questions asked of each student. The value of 1 indicates an extremely liberal response, 4 is midway, and 7 indicates an extremely conservative response. Appropriate values lie in between. The task is to form a crosstabulation between gender and liberalism/conservatism ratings.

To do this, choose the CROSSTABS subprogram. Request Cramer's V as the statistic to assess the degree of association between these dimensions. For the present, you do not need to practice with all of the statistical options, but by all means practice with those with which you are familiar.

Create a file for these data, recalling that the position of the SAVE FILE instruction is fixed. Also recall that in creating a permanent file, additional SCL is necessary (ask your consultant). Note in your reading of the CROSSTABS subprogram that the INTEGER versus GENERAL mode is much the same as the INTEGER versus GENERAL mode in the FREQUENCIES subprogram.

Answer:

```
RUN NAME          CROSSTAB OF GENDER BY LIB-CON RATING POL ATTITUDES STUDY
FILE NAME         POLATTIT,POLITICAL ATTITUDE SHIFTS OF UNDERGRADUATES
VARIABLE LIST     ID,NAME,SAT,STATE,LOCALITY,AGEF,AGEM,MAJOR,SEX,AUTHF,AUTHM,
                  AUTHS,INTERID,FQ1 TO FQ9,MQ1 TO MQ9,CLASS,AGESTU,SQ1 TO SQ9
INPUT MEDIUM      CARD
N OF CASES        6
INPUT FORMAT      FIXED(1A3,1X,2A4,1X,F3.2,1X,F2.0,F3.0,1X,2F2.0,F3.0,F1.0,3F
                  2.0,1X,F2.0,9F1.0,1X,9F1.0,1X,F1.0,F2.0,9F1.0)
MISSING VALUES    SAT(999)/STATE,LOCALITY,AGEF,AGEM,AUTHF,AUTHM,AUTHS,AGESTU(99)
                  /SEX,INTERID,CLASS,FQ1 TO FQ9,MQ1 TO MQ9(9),SQ1 TO SQ9(9),
                  MAJOR(0)
VAR LABELS        (same as in Exercise 2)
VALUE LABELS      STATE(1)PA,(2)NJ,(3)DE,(4)MD,(5)NY,(6)CN,(7)OH,(8)WV,
                  (9)VA,(10)MA,(11)DC,(12)UNKNOWN/SEX(1)MALE,(2)FEMALE/
                  CLASS(1)FR,(2)SO,(3)JR,(4)SR
CROSSTABS         VARIABLES=SEX(1,2),SQ9(1,7)/TABLES=SEX BY SQ9
OPTIONS           9
STATISTICS        2
SAVE FILE
FINISH
```

The inclusion of a FILE NAME instruction is the first difference between this solution and that of Exercise 2. This instruction is generally used only once; thereafter a command such as GET FILE POLATTIT would be used. Note that a SAVE FILE instruction is also required when the file is first created. Beyond these differences, the instructions are as in Exercise 2, until we come to the task definition cards.

We can see from the CROSSTABS instruction that the INTEGER mode was chosen. This is entirely appropriate, since neither variable contains decimal values. Decimal and alphanumeric values in any variable of concern require the GENERAL mode. The distinction is made because of the speed and the storage savings possible with known integer data.

Complications arise when crosstabulation tables are produced in relation to another variable. For example, it might be of interest to examine SEX BY SQ9 for each state on the grounds that there are prevailing regional political differences. In the GENERAL mode we would write:

```
CROSSTABS         TABLES=SEX BY SQ9 BY STATE
```

The limit is 20 such lists, and of course the CROSSTABS procedure could be requested several times to accomplish the same result. Implied lists (e.g., Q1 TO Q9) may be used in both the VARIABLES and the TABLES portions.

Even in a very modest survey such as the one presented here, a large number of tabulations could be produced. The best advice for the beginner is to be conservative in requesting only those tables that are of central interest. Option 9 can be of significant help when large numbers of tables are produced. Of all the options available, number 9 is of greatest importance. This option creates an index of tables, enabling the user to find the page number of any given table. In this practice exercise, however, option 9 is trivial.

The STATISTICS instruction requests Cramer's V as specified. If the specification SEX BY SQ9 had resulted in a 2 X 2 table, then phi would have been calculated instead. It is a dubious procedure to request all statistics, for many of the statistics are totally inappropriate.

CROSSTABS is often used in conjunction with variables that are derived rather than with those that are original. The topic of variable and file modification is taken up in the next even-numbered exercise.

─────── EXERCISE 5 ═══

After obtaining a frequency tabulation for the data in Exercise 3, the next step is to test whether or not the treatment groups differ. This naturally calls for an analysis of variance (ANOVA). However, a cautious researcher might well be concerned about male/female differences with respect to the dependent measure. Generally, such differences either become part of the experiment or else they are tested for discrepancies. If no discrepancies are found, the two groups are merged into one. In this exercise, we will begin with the preliminary phase in order to demonstrate the use of the SPSS T-TEST procedure. Also, assume that it is desirable to create a permanent file for further testing.

Do a two-tailed independent-samples t-test for differences between gender (SEX) for each of the three testing conditions. As before, the dependent measure is the grade on the final exam. However, since there are three groups (and not one), we need a method of informing SPSS of this fact. The standard one is the data selection instruction SELECT IF. Alternatively, the data could be given a subfile structure and the analysis performed on each subfile. A third alternative—and one that does not introduce a new topic—is the judicious use of the MISSING VALUES instruction. By writing three sets of T-TEST instructions and preceding each with a MISSING VALUES statement that declares the two unwanted groups as missing, we accomplish the same thing as with the first two procedures. Use this latter procedure for this exercise. Be careful not to specify an option that would defeat the selection effect.

As a final point, use the EDIT feature as a means of checking for correct PCL instructions.

Answer:

```
EDIT
RUN NAME         CHECK FOR SEX DIFFERENCES IN TEST ANXIETY
FILE NAME        TESTANX
VARIABLE LIST    ID,AGE,SATMATH,SATVERB,SEX,CLASS,MAJOR,GPA,TANXGRP,TESTCOND,
                 FINAL,TANXS
INPUT MEDIUM     CARD
N OF CASES       18
INPUT FORMAT     FIXED(1A3,1X,F2.0,1X,2F3.0,2F1.0,1X,F3.0,1X,F3.2,2F1.0,1X,2F2.0)
VAR LABELS       GPA,GRADE POINT AVERAGE/TANXGRP,TEST ANXIETY GROUP/TESTCOND,
                 TESTING CONDITION/FINAL,FINAL-EXAM GRADE/TANXS,TEST ANXIETY SCORE
VALUE LABELS     SEX(1)MALE(etc., same as in Exercise 3)
MISSING VALUES   AGE TANX(99)/SATMATH SATVERB(000,999)/SEX CLASS TANXGRP(9)/
                 MAJOR GPA(999)/FINAL(00)/TESTCOND(2,3,9)
T-TEST           GROUPS=SEX/VARIABLES=FINAL
READ INPUT DATA
MISSING VALUES   TESTCOND(1,3,9)
T-TEST           GROUPS=SEX/VARIABLES=FINAL
MISSING VALUES   TESTCOND(1,2,9)
T-TEST           GROUPS=SEX/VARIABLES=FINAL
SAVE FILE
FINISH
```

The first thing to note about this answer is that the data are missing. They normally would be located immediately after the READ INPUT DATA instruction.

After the RUN NAME command, we can see that the name chosen for the SPSS file is TESTANX. Any name is acceptable as long as it conforms to the usual naming conventions. By naming a file, we do not automatically save it. The SAVE FILE instruction accomplishes this if it is keyed before the FINISH command. No other position of the SAVE FILE instruction is allowed.

All instructions from VARIABLE LIST to VALUE LABELS are the same as in Exercise 4. The MISSING VALUES function is the same except for the position of TESTCOND. As can be observed, it has been moved merely to improve clarity, and values 2 and 3 have been added to the standard missing value of 9. The result is the elimination of conditions 2 and 3 from the immediate analysis; they are not removed from the file, however.

In the first T-TEST instruction, we can see that GROUPS is set equal to SEX, and the variable to be analyzed is FINAL. More than one dependent variable may be specified if desired. Reading through the remaining instructions, we can see that a new set of missing values is declared and that the same T-TEST instruction is given. Thus, a male-versus-female test is made for each of the test condition groups.

The T-TEST instruction may take several forms. The values that determine the two groups may be indicated as in the following:

```
T-TEST          GROUPS=SEX(1,2)/VARIABLE=FINAL
```

The answer above is a special case. If the values are 1 and 2, then they are not required; that is, these values need not be specified for GROUPS. Yet another alternative is to select a single value. Group 1 is composed of those cases greater than or equal to the selected value, and group 2 is composed of the remaining cases. Groups can also be formed by selecting the first n cases and the next n cases, as in the following:

```
T-TEST          GROUPS=9,9/VARIABLES=FINAL
```

This selects the first 9 cases as group 1 and the next 9 as group 2. Extra cases are not analyzed. These techniques provide the user with a wide variety of tools and of case-selection procedures. See the "Pro Tips and Techniques" section for additional details.

7.3.3 Practice with the SCATTERGRAM and NPAR TESTS Subprograms

These exercises have four objectives: to establish familiarity with the SCATTERGRAM subprogram, to practice with multiple-task definitions, to introduce the topic of data modification, and to demonstrate Pathway III.

═══════ EXERCISE 6 ═══════════════════════════════════

To accomplish the first objective, that of familiarity, you should read the SPSS description of SCATTERGRAM and then write the instructions to obtain plots of the fathers', mothers', and students' authoritarian scores versus the students' ages. Consider the scores to be on an interval scale. Also, obtain a Pearson's r statistic and a two-tailed test of significance.

The second objective, practice with multiple-task definitions, can be achieved by obtaining another complete scattergram run (include a second set of SCATTER-GRAM, OPTIONS, and STATISTICS instructions) of the authoritarian scores versus the students' class. Note that in this case Pearson's r and other such statistics should be ignored because CLASS is not a continuous, normally distributed variable.

The third objective, introducing data modification, is more complex. The goal is to reduce the three sets of conservatism/liberalism questions to a single variable or scale score. Obtain the new variable by simply adding the questions. (The advisability of doing this is beyond the scope of the current discussion.) The operation needed is generally referred to as recoding, or transformation. In SPSS, this operation is accomplished through the data-modification instruction. Since data modification in SPSS is a lengthy topic, it is suggested that you learn the COMPUTE instruction. This is not to say that other instructions, such as RECODE, are useless. They will be taken up in the next exercise. (*Hint:* In data modification, individual variables may be treated as if they were ordinary algebraic quantities.)

Use the newly computed variables to obtain scattergrams of the father, mother, and student versus each other on the conservatism/liberalism dimension. Assume that a file named POLATTIT has already been established.

Answer:

```
RUN NAME       PLOTS,AUTHORITARIANISM,CONSERVATISM-LIBERALISM,AGE,CLASS
GET FILE       POLATTIT
COMPUTE        CONLIBF=FQ1+FQ2+FQ3+FQ4+FQ5+FQ6+FQ7+FQ8+FQ9
COMPUTE        CONLIBM=MQ1+MQ2+MQ3+MQ4+MQ5+MQ6+MQ7+MQ8+MQ9
COMPUTE        CONLIBS=SQ1+SQ2+SQ3+SQ4+SQ5+SQ6+SQ7+SQ8+SQ9
SCATTERGRAM    AGE WITH AUTHF,AUTHM,AUTHS
OPTION         6
STATISTICS     1,3
SCATTERGRAM    CLASS WITH AUTHF,AUTHM,AUTHS
OPTION         7
STATISTICS     ALL
VAR LABELS     CONLIBF,CONSERVATISM-LIBERALISM OF FATHER/CONLIBM,CONSERVATISM-
               LIBERALISM OF MOTHER/CONLIBS,CONSERVATISM-LIBERALISM OF STUDENT
SCATTERGRAM    CONLIBF(9,81) TO CONLIBM(9,81)
STATISTICS     ALL
FINISH
```

The most obvious conclusion we can draw from this answer is that, once established, SPSS system files can be used to make the entire process much easier. New VAR LABELS, MISSING VALUES, and VALUE LABELS instructions could have been stated after the GET FILE command if there had been a need.[1] The central point still remains that, in Pathway III, a great many instructions are no longer required because their products are permanent parts of the system file.

In the first SCATTERGRAM instruction, we can see the use of the qualifier WITH. SPSS refers to these qualifiers as keywords. In this instance, three plots would be generated: AGE by AUTHF, AGE by AUTHM, and AGE by AUTHS. The usual variable rules apply; i.e., implied lists are acceptable as SCATTERGRAM instructions. For example, if we desired the 9 student questions plotted against the 9 father questions, we would type this instruction:

```
SCATTERGRAM    SQ1 TO SQ9 WITH FQ1 TO FQ9
```

The 9 student questions would each become the vertical axis of 9 separate plots, for a total of 81. Once again, the power of SPSS must be kept under control. The keyword TO may also be used alone, as in the following:

```
SCATTERGRAM    SQ1 TO SQ9
```

which produces all 36 combinations of two-way plots.

---

[1] The initial labels are a permanent part of the file and are available at each use.

The instruction OPTION 6 in combination with the 3 in the STATISTICS instruction produces the requested two-tailed test of the Pearson *r*. Both of these instructions function just as they did in previous exercises and subprograms.

The next SCATTERGRAM instruction is similar to the preceding one. However, it is important to note that if either of these instructions had referred to variables derived from the COMPUTE instruction, an error condition would have resulted. All data modifications must occur before they can be used in a subprogram request. In this instance, COMPUTE operations could appear anywhere between the GET FILE and the third SCATTERGRAM instruction because the COMPUTE commands will generate new variables and will not be needed until the last subprogram. Data modification that changes old variables is an irrevocable step; the old values are lost. See the "Pro Tips and Techniques" section for solutions.

The three COMPUTE instructions illustrate several features. Continuation of similar instructions with a slash is not allowed; i.e., the three COMPUTE functions could not be combined. However, multiple algebraic operations are possible. Parentheses may be used whenever the notation becomes ambiguous. For example, suppose we wish to add the first 3 student questions and then divide them by 2. We would have:

```
SCATTERGRAM     NEWSVAR=SQ1+SQ2+SQ3/2
```

or:

```
SCATTERGRAM     NEWSVAR=(SQ1+SQ2+SQ3)/2
```

The first calculation is ambiguous in that it is not clear if the additions are to be done before the division by 2, or if one should divide SQ3 by 2 and then add the results to SQ1 and SQ2. The second COMPUTE instruction is unambiguous: add and then divide. Actually SPSS interprets such arithmetic instructions in a hierarchical fashion that becomes quite complex in use. Remember to use parentheses. If the above examples are still unclear, insert the values 2, 7, and 4 into the equations. The results are 6.5 in the first case and 11 in the second.

Calculated variables are not limited to arithmetic calculations; trigonometric and algebraic calculations are also possible. When logical qualification is required (for example, "Include only if subjects are older than forty years of age"), the keyword IF is used (see Exercise 7).

Note that CONLIBF, CONLIBM, and CONLIBS would not need to be recalculated in an analysis at a later date. They become permanent parts of the file. (Preceding a COMPUTE command with an asterisk designates the calculated variable as a temporary one that is not retained in the permanent file.)

In the final SCATTERGRAM, each variable has been qualified by having its lowest and highest values defined. This is useful, since the resulting plots are more readable and waste less space. However, as stated, the CONLIBM function would go from lowest to highest and not from 9 to 81. The range must be set for each variable. If the actual lowest calculated value of CONLIBM was 5, the three plots would not be at all compatible.

We can see from Exercise 6 that a great deal can be packed into a single SPSS analysis. Beginners should add new instructions cautiously because introducing a large number of changes makes it next to impossible to determine what caused an error. And experience indicates that errors are inevitable.

===== EXERCISE 7 =====

Read the SPSS description of NPAR TESTS (nonparametric tests), and then write instructions to test the hypothesis that the three test-anxiety groups are from the same population. Unlike most SPSS procedures, NPAR TESTS has 13 subsections. The choice in this case is the Mann–Whitney U test. Conceptually, it is similar to a t-test but does not have parametric assumptions. Obtain three separate Mann–Whitney U's that compare the three separate text-anxiety groups with each other. Accomplish this through separate task definitions. The dependent measure is the final exam grade.

As a second task program, obtain a similar Mann–Whitney U with the same dependent measure, but use SAT math and verbal scores to define the groups. Do this by determining which is higher—math or verbal. Thus, group 1 has high math scores and group 2 has high verbal scores. For the present purposes, do not attempt to resolve tie scores.

For the above tasks, assume that the data have been saved in a file named TESTANX. Also obtain the standard descriptive statistics, means, and so on.

Answer:

```
RUN NAME        TEST OF PREEXISTING DIFFERENCES ON TEST ANXIETY AND SAT
GET FILE        TESTANX
IF              (SATMATH-SATVERB GE 1)SATGROUP=1
IF              (SATMATH-SATVERB LT 1)SATGROUP=2
VAR LABELS      SATGROUP,GROUP ASSIGNMENT
VALUE LABELS    SATGROUP(1)HIGH MATH SCORE(2)HIGH VERBAL SCORE
NPAR TESTS      M-W=FINAL BY TANXGRP(1,2)
STATISTICS      1
NPAR TESTS      M-W=FINAL BY TANXGRP(1,3)
NPAR TESTS      M-W=FINAL BY TANXGRP(2,3)
NPAR TESTS      M-W=FINAL BY SATGROUP(1,2)
STATISTICS      1
SAVE FILE
FINISH
```

The most obvious aspect of this answer is the absence of many of the customary preliminary statements. Such information is not absent but is rather a routine part of the file TESTANX. Thus, the time and trouble it takes to establish a file is certainly worth the effort when repeated analyses are expected.

The four instructions after GET FILE serve to create new variables that are based on the relationship of the SAT scores. Since this relationship pertains to the second half of the exercise, the discussion is delayed until then.

The first NPAR TESTS instruction clearly requests a Mann–Whitney U (M-W) with the final-exam grade as the dependent variable and with the test-anxiety group as the independent variable. Those cases that are coded 1 become group 1, and those coded 2 become group 2. The values pertain to the codes present in the variable itself; for example, this instruction:

```
NPAR TESTS       M-W=FINAL BY TANXGRP(273,17)
```

would form group 1 from all cases coded 273, and would form group 2 from all cases coded 17. If this were done with age as a variable, only two ages–not two ranges of ages–would be analyzed.

The STATISTICS instruction requests the standard descriptive statistics and is the only request possible.

The next two NPAR TESTS requests are essentially the same as the first except that they make comparisons between different subgroups. Other available analyses could be requested in a similar manner. In other words, all thirteen nonparametric tests are available within a single run.

The last NPAR TESTS request is for the final-exam grade (FINAL) by the new variable SATGROUP, which is determined from the data-modification instruction above. IF instructions can be quite complex, but they all have the same general form: "If this and this are true, then x equals y." The present example would be translated as follows: "If SATMATH minus SATVERB is greater than or equal to 1, then the new variable SATGROUP is equal to 1." The second IF instruction is similar except that it determines if the verbal score is greater than the math score. This appears to be a contradiction, for the logical expression is "is less than" (LT) and not "is greater than." This is an example of the situation in which there are several logical ways of making the same assignment. We could have written this IF instruction:

```
IF          (SATVERB-SATMATH GT 1)SATGROUP=1
```

and accomplished exactly the same thing. The expression "greater than or equal to" (GE) could not have been used because tie scores would be assigned both value 1 and value 2. The effect of these statements is to assign a new variable to each case, coded 1 or 2 depending upon the respective value of the SATVERB and the SATMATH variables.

The IF form of data modification may be combined with other forms in a set of complex instructions. Indeed, the IF itself may take on very involved forms. It may seem obvious, but the beginner should proceed with caution with complex

IF instructions. In this exercise, for example, ties are handled by placing them in group 1. A resolution of this would be the following:

```
IF              (SATMATH-SATVERB GE 1)SATGROUP=1
IF              (SATMATH-SATVERB LT 0)SATGROUP=2
IF              (SATMATH-SATVERB EQ 0)SATGROUP=9
MISSING VALUES SATGROUP(9)
```

This solution eliminates tie scores from consideration. If a statistic other than the Mann–Whitney U were of interest, then the tie scores could be retained and analyzed.

Once computed, such assignments become a permanent part of the file (if the file is saved, of course). Temporary assignments can be made by placing an asterisk in front of the IF instruction. Temporary assignments are not saved. The location for all permanent data-modification statements is before the first task assignment, but they need not be in any specific order. The exception is that permanent modifications must precede temporary ones, and temporary ones may also be located after the first procedure instruction. (See the "Pro Tips and Techniques" section.)

The VAR LABELS and VALUE LABELS instructions serve the obvious purpose of labeling the newly computed variables. These instructions should not precede the determination of the new variables, however.

The last NPAR TESTS and STATISTICS instructions offer no surprises. Remember, nonparametric tests do not require normal distributions or interval data. Test-anxiety measures often cannot meet these requirements. Thus, in the test-anxiety group analyses above, nonparametric tests are preferred. A further word of caution —although many individual tests may be performed on the same data with nonparametric tests, it is not advisable to do so in many research situations.

7.3.4   Practice with the REGRESSION and ANOVA Subprograms

Practice with the REGRESSION and ANOVA subprograms and use of the SPSS RECODE facility are the goals of the following exercises. RECODE is an important facility that warrants practice; complete both exercises.

══════ EXERCISE 8 ══════

To demonstrate the power of SPSS in restructuring variables, recode the college major variable from the conservatism/liberalism study into one containing only three values: liberal arts majors, science majors, and professional majors (business and engineering). Consider numeric scores 1 through 57 as liberal arts majors, 60 through 92 as science majors, and 100 through 135 as professional majors. Note that the gaps in this sequence should be considered as missing data, although ostensibly there would be no one with a major having a value of 59 or 97.

For the regression, consider the students' conservatism/liberalism score as the dependent, or predicted, variable. Use the remaining data as independent variables except for ID, name, locality, and interviewer ID. (The advisability of using these particular variables or any subset of them is a statistical decision beyond the scope of this discussion.) As to the kind of analysis, choose a forward stepwise regression that is limited to the best eight predictors and is without a hierarchical structure. Take the default for RESID, and do not choose any of the available options until you are quite familiar with this subprogram. Also, it is best to print the correlation matrix that is available through the STATISTICS instruction.

If the computation of conservatism/liberalism scores had not been done prior to the regression analysis, it would be necessary to do it here. Even though these calculations were a part of Exercise 6, they should be redone here for clarity and practice. Assume a file named POLATTIT has been established as in Exercise 4.

Answer:

```
RUN NAME        REGRESSION ON STUDENTS CONSERVATISM-LIBERALISM SCORE (CONLIBS)
GET FILE        POLATTIT
COMPUTE         CONLIBF=FQ1+FQ2+FQ3+FQ4+FQ5+FQ6+FQ7+FQ8+FQ9
COMPUTE         CONLIBM=MQ1+MQ2+MQ3+MQ4+MQ5+MQ6+MQ7+MQ8+MQ9
COMPUTE         CONLIBS=SQ1+SQ2+SQ3+SQ4+SQ5+SQ6+SQ7+SQ8+SQ9
RECODE          MAJOR(1 THRU 57=1) (60 THRU 92=2) (100 THRU 135=3)
VALUE LABELS    MAJOR(1)ARTS MAJOR(2)SCIENCE MAJOR(3)BUSINESS OR ENGINEER
MISSING VALUES  MAJOR(58,59,93 THRU 99)
REGRESSION      VARIABLES=CONLIBS,CONLIBF,CONLIBM,SAT,STATE,AGEF TO AUTHS,CLASS,
                AGESTU/REGRESSION=CONLIBS(8) WITH CONLIBF,CONLIBM,SAT,STATE,
                AGEF TO AUTHS,CLASS,AGESTU(1)
STATISTICS      1
FINISH
```

The GET FILE and COMPUTE instructions are the same as in Exercise 7. Next, we can see that the RECODE facility operates much the same as other SPSS instructions. If the variable MAJOR were continuous from 1 to 135, a multiple specification would still be necessary. A variable may be recoded into as many classifications as are appropriate. Combinations of the COMPUTE and RECODE requests are acceptable, and if a temporary recoding is desirable, the asterisk convention can be used.

The VALUE LABELS instruction following the RECODE command is not necessary. But, as always, it is a great convenience because it is very easy to forget that variables were modified in a certain analysis.

It is also not necessary to specify the breaks in the variable MAJOR as missing, but this procedure does provide excellent protection against incorrectly coded data. This is an especially good idea when most variables use 9, 99, or some similar number as the missing code and when the variable in question uses a different code (0 in this case).

In the task definition instructions, we can see that all variables considered must be listed, followed by the specific regression design. The 8 that follows the dependent measure (CONLIBS) restricts the analysis to the best eight predictors. The default is all predictors. The F for significance and tolerance would have been included with the 8 if other than default values had been desired.

In a similar fashion, the 1 that follows the variable AGESTU(1) specifies a forward-stepping regression without a hierarchical structure. There are many different regression designs possible; see your SPSS manual for details. We can also note from the task definition instructions that implied lists are acceptable. If each variable in a list is to be treated differently, then implied lists may not be used.

========= **EXERCISE 9** =========

The objectives of this exercise are to practice with the ANOVA subprogram and to use the RECODE facility. To demonstrate the power of SPSS in restricting variables, recode the grade point average (GPA) variable from the test-anxiety study (Exercise 3) into six categories. If we assume that the basis is a four-point GPA system, then the first level would be grade point averages from 1.00 to 1.49, the second level would be from 1.50 to 1.99, and the successive intervals would be from 2.00 to 2.49, from 2.50 to 2.99, from 3.00 to 3.49, and from 3.50 to 4.00. In SPSS terms, this recoding changes a metric variable (the GPA) into a nonmetric, or categorized, variable.

For the ANOVA operation, assume that the final-exam grade is the dependent variable and that there are three independent variables: the testing condition, the gender, and the GPA modified as above. Recall that testing involved: a timed exam, an unlimited time exam, and a timed exam accompanied by "mood" music. Since the gender variable has two levels, or variations, we have a three-factor study; the testing condition has three variations, the gender has two, and the GPA has six.

Write the SPSS instructions for the above study using the classic experimental approach. Do not choose any of the options or statistics because these are concerned with the nonclassical approaches and covariance analysis. Note that the SPSS ANOVA subprogram does not allow either random or mixed-effects models. Unequal cell sizes are acceptable, however.

Assume that a file named TESTANX has been previously established. In your SPSS reading, the following topics are not essential to ANOVA and may be skipped: covariance analysis, hierarchical and regression approaches, and multiple-classification analyses (MCA). However, be sure to read about these subjects when you are ready for them. Indeed, it is advisable to attempt a second analysis, with the GPA as a metric (noncategorized) covariate.

Answer:

```
RUN NAME      ANOVA TEST CONDITIONS, GENDER, GPA ON FINAL EXAM GRADE
GET FILE      TESTANX
RECODE        GPA(1.00 THRU 1.49=1) (1.49 THRU 1.99=2) (1.99 THRU 2.49=3)
              (2.49 THRU 2.99=4) (2.99 THRU 3.49=5) (3.49 THRU 4.0=6)
```

```
VALUE LABELS    GPA(1)1.00 THRU 1.49(2)1.50 THRU 1.99(3)2.00 THRU 2.49(4)2.50
                THRU 2.99(5)3.00 THRU 3.49(6)3.50 THRU 4.00
MISSING VALUES  GPA(0.00 THRU 0.99)
ANOVA           FINAL BY TESTCOND(1,3)SEX(1,2)GPA(1,6)
SAVE FILE
```

The RUN NAME and GET FILE instructions do not present new information. However, the RECODE instruction is new, and we can observe that it operates in much the same way as do other SPSS instructions. Each variable is listed, with the nature of the modification following in parentheses. A variable may be modified into as many classifications as are appropriate. When a series of variables or an implied list of variables is to be recoded in the same way, a single set of specifications will suffice. For example, we could have:

```
RECODE          ITEM1,ITEM8,ITEM17(1 THRU 2=1) (3 THRU 5=2) (6 THRU 7=3)
VALUE LABELS    ITEM1,ITEM8,ITEM17(1)LOW(2)MED(3)HIGH
```

and:

```
RECODE          ITEM1 TO ITEM91(1 THRU 3=1) (4 THRU 6=2) (7 THRU 9=3)
VALUE LABELS    ITEM1 TO ITEM91(1)LOW(2)MED(3)HIGH
```

The TO convention may also be used when variables are adjacent in the VARIABLE LIST instruction. If temporary modifications are desired, simply precede each RECODE request with an asterisk. Recoded variables are permanent only for the current SPSS task. However, if a SAVE FILE instruction is included as it is in this example, then the file itself will be modified as instructed. The instructions marked *RECODE are temporary even if a SAVE FILE command is issued.

Note that in this answer, the values of the GPA actually overlap; i.e., the second value begins with 1.49 and not with 1.50. This is a common precaution because the internal representation of a number often is not perfect. Thus, if by rounding errors 1.49 is internally presented as 1.4900001, it would not be recoded because it lies in the gap between 1.4900000 and 1.5000000 (assuming accuracy to nine digits). By overlapping the values, we ensure against data loss.

In the task-definition instruction, we can see that the dependent variable is stated first, followed by the independent variables. Each of the independent variables includes in parentheses its minimum and maximum values, in effect defining the levels of each factor. (See the "Pro Tips and Techniques" section for more information on this.)

The ANOVA subprogram is quite complex and provides a wide variety of analyses, particularly those that are more common in sociology and in political science (rather than in psychology). One example of the capabilities of this facility is found in the task suggested at the end of the instructions for this exercise. In this covariance analysis, the GPA variable generally would not be recoded into a discrete classification but would be left as a continuous, metric variable. Thus, the RECODE, the VALUE LABELS, and the MISSING VALUES operations in this answer would not be included, and the task would read as follows:

ANOVA   FINAL BY TESTCOND(1,3)SEX(1,2) WITH GPA

This is a two-way analysis with one covariate.

There are far too many variations to be printed here. However, beginners should assure themselves that the assumptions built into this subprogram are indeed the proper ones for their needs. It is one thing to get the SPSS package to function; it is another thing to meet assumptions.

7.3.5   Practice with the DATA LIST Option

Since one data list is more or less similar to another, it will not be necessary to present separate exercises in this section. All exercises so far have utilized Pathways I and III. In the final exercise, we will briefly examine Pathway II—the path of the DATA LIST input method. To better contrast methods, we will not introduce a new subprogram; rather, we will redo Exercise 2. Think of the DATA LIST method as the VARIABLE LIST and the INPUT FORMAT methods merged into one.

════════ **EXERCISE 10** ════════════════════════════

Rewrite Exercise 2 using the DATA LIST method (Pathway II) with the data on cards. There is no particular advantage to repeating the MISSING VALUES, VAR LABELS, and VALUE LABELS instructions. Obtain frequencies on all variables for which it is logical. Use the EDIT instruction only if it seems appropriate to you. Refer to Section 7.2.3 to review the DATA LIST option.

Answer:

```
RUN NAME        DATA LIST INPUT OF POLITICAL ATTITUDE SHIFTS STUDY
DATA LIST       FIXED(1)/1 ID 1-3 (A) NAME 5-12 (A)SAT 14-16(2)STATE 18-19
                LOCALITY 20-21 AGEF,AGEM 23-26 MAJOR 27-29 SEX 30 AUTHF,AUTHM,
                AUTHS 31-36 INTERID 38 FQ1 TO FQ9 39-47 MQ1 TO MQ9 49-57
                CLASS 59 AGE 60-61 SQ1 TO SQ9 62-70
INPUT MEDIUM    CARD
N OF CASES      6
FREQUENCIES     INTEGER=SAT(100,99) STATE TO AGEM,AUTHF TO AUTHS,AGESTU(01,
                99) FQ1 TO MQ9,SEX,CLASS,SQ1 TO SQ9(1,9)
OPTIONS         7
STATISTICS      ALL
READ INPUT DATA
```

The only topic in this answer that has not yet been treated is that of DATA LIST itself. To this end, we will decode the instructions step by step. The alternative to the FIXED mode is the BINARY mode. The average user is unlikely to have data in a binary format, so this alternative is strictly a "pro" technique and need only be mentioned here.

Immediately after the word FIXED is (1), indicating that there is only one record (line or card) per case. Strictly speaking, this term is unnecessary, since 1 is the number that is assumed. The number of possible records per case is not infinite but is respectably large. In actuality, the number is determined by the number of variables (see Section 7.4).

Next we find the description of each card or line. The description begins with a slash and would end with one if the case continued onto the next card or line. Following the slash, further identification is made with an identifier. The 1 immediately after the slash identifies what follows as the specification for the first record. Then each variable is identified in sequence. If the ID were on the first record, the name on the second record, and the remaining variables on the third record, we would have:

```
DATA LIST        FIXED(3)/1 ID 1-3(A)/2 NAME 5-12(A)/3 SAT 14-16 . . .
```

This could be the case only if these variables were in the indicated columns, which is unlikely in a real analysis. The point is that if a DATA LIST instruction begins with record three, then records one and two are unavailable to the analysis. Thus, this instruction:

```
DATA LIST        FIXED(7)/7 ID 1-3(A) . . .
```

precludes variables being read from records one through six. Within each record, variables may be read in any order. Indeed, this creates a powerful tool for reordering variables. If it were necessary that all questions be grouped together, we would have the following:

```
DATA LIST        FIXED(1)/1 ID 1-3(A) NAME 5-12(A) SAT 14-16(2) STATE 18-19
                 LOCALITY 20-21 AGEF,AGEM 23-26 MAJOR 27-29 SEX 30 AUTHF,
                 AUTHM,AUTHS 31-36 INTERID 38 CLASS 59 AGE 60-61 FQ1 TO FQ9
                 39-47 MQ1 TO MQ9 49-57 SQ1 TO SQ9 62-70
```

Any order within a card is acceptable. In the initial or in the "clean-up" phases of an analysis, the original order should be maintained.

From this example, we can see that the TO convention applies. However, the techniques that work for VAR LABELS and MISSING VALUES do not apply. Recall that these techniques depend on the VARIABLE LIST instruction to establish variable adjacency and thus cannot be used here.

The specifications for each variable are direct. The column locator of each variable is stated, followed when necessary by a decimal designator or an alphabetic designator. The (A) following ID and NAME labels them as alphabetic. The (2) after SAT indicates two decimal places. When decimals are punched, special designators are not required.

Column locators may be single (as in SEX 30) or they may be any string of adjacent columns (for example, ID 1-3, STATE 18-19, and MQ1 TO MQ9 49-57). Obviously, specifications greater than 80 columns are disallowed when using cards.

Many terminal systems will allow longer lines, which may or may not be correctly translated by SPSS. Ask your consultant about the system you are using.

The DATA LIST option (Pathway II) does allow for skipping of variables by exclusion. In the example, column 13 is not specified with any of the variables and thus is effectively ignored. This applies to any number or combination of columns.

7.4   Pro Tips and Techniques for SPSS

This section is not really aimed at the professional (who probably would not need to use this book). It is intended for the moderately experienced SPSS user who can become more polished by reviewing the four major problem areas and learning the solutions a pro would use. The most common errors are the use of abbreviations and incorrect plurals, the incorrect ordering of instructions, the misuse of SPSS syntax, and incorrect data modification.

7.4.1   Abbreviations and Plurals

Unlike the other analysis packages discussed in this book, SPSS does not allow abbreviations. Truncation, which can be seen as an exception, refers to the fact that only the first eight characters of a control word are interpreted; for example, VALUE LA could be used instead of VALUE LABELS. However, it is probable that this eight-character rule produces more confusion than benefit, and therefore it is not recommended.

Plurals can also be somewhat confusing. In general, you should spell as the SPSS manual suggests. MISSING VALUE will be interpreted correctly, whereas ANOVAS will produce an error condition. With any control word greater than eight characters, plurals do not matter; but with control words of less than eight characters, plurals cannot be used.

It is best to write all instructions as they are specified in the manual. However, there are some additional exceptions in the SCSS package.

7.4.2   Order of Program Control Language (PCL) Instructions

The order of program control language (PCL) instructions is very important in SPSS, but there is no absolute rule or accepted order. The manuals all state the minimum order. However, the actual logic of the instructions allows for certain variations.

Instructions such as MISSING VALUES must follow any data transformations to which they apply. However, such data-defining instructions could precede transformations if the instructions refer only to the input data. If data-defining instructions are written that refer to both input data and transformation variables, the requests must follow the transformations.

Temporary transformations and selection instructions may be placed with the individual task (subprogram) to which they apply. Alternatively, they may come before all task instructions. In this case, one must be cautious, for unintended results may be obtained. When a variable must be modified in one way for a given analysis and in another way for a later task, the modifications must be paired with their respective tasks. For example, in studying men and women separately, the grouping of temporary SELECT IF instructions will cause them to apply to all subsequent analyses. Consider the following:

```
*SELECT IF      (SEX EQ 1)
*SELECT IF      (SEX EQ 2)
FREQUENCIES     GENERAL=SEX
FREQUENCIES     GENERAL=SEX
```

The effect of the above would be to produce two identical frequency analyses on the variable SEX instead of producing one analysis for men and one for women.

SPSS interprets multiple *SELECT IF commands by connecting them with a logical OR. Thus, this instruction:

```
*SELECT IF      (SEX EQ 1 OR SEX EQ 2)
```

produces the same result as that from the two instructions above. Multiple SELECT IF operations function in the same way. The solution is to keep modifications with their respective analyses. Be cautious with combinations of permanent and temporary SELECT IF functions. SPSS interprets these as if they were connected with a logical AND.

To further complicate the issue, SPSS has four versions that each interpret multiple SELECT IF commands differently. In Version 6, they are connected by a logical OR; i.e., cases are selected that satisfy either condition. In Versions 7, 8, and 9, these operations are connected by the logical AND. In the preceding example, Versions 7, 8, and 9 would select no cases, for none would have both a code 1 and a code 2.

The order, or precedence, of all modifications is not fixed. A COMPUTE command may precede or follow a SELECT IF command, and so forth. What is important is the meaning of the resulting statement; it must make sense logically. The wide ramifications of this warrant a separate discussion, which is presented in Section 7.4.4.

Overall, the order of SPSS instructions is fairly strict and must be adhered to. The manuals that accompany all versions give detailed information on this issue. There are two special rules to note: the EDIT facility does not check for instruction order; and the SAMPLE data-modification instructions are executed in sequence in Versions 7, 8, and 9 but are always first in Version 6. A simple violation of the general constraints on SPSS order is found in Version 9; it does not require a FINISH instruction. In reality, SPSS will supply the statement when necessary.

7.4.3   Syntax of SPSS

Experience indicates that syntax is one of the areas of SPSS in which users make many errors. This refers to situations in which COMPUTE, IF, and RECODE are used in conjunction to create new variables and to modify old variables. SELECT IF presents the same problems. When only one operator is present, there is not much trouble; however, it is not uncommon to find 25 to 30 operators working at one time. Generally, it is the combinations of instructions that produce problems, coupled with the fact that they do not all have the same priority or precedence. To illustrate, recall the survey of political attitude shifts among college students, and suppose that the researcher wishes to determine the frequency of students whose father's and mother's authoritarian scores combined are over 29 (high) and whose own score is under 16 (low). The essentials are as follows:

```
IF              ((AUTHF AND AUTHM) GE 30 AND AUTHS LE 15)ATSHIFT=1
FREQUENCIES     INTEGER=ATSHIFT(0,1)
```

The order of evaluation is the critical element in this instruction. The order is as follows: the inner parentheses, the GE (greater than or equal to), the LE (less than or equal to), and lastly, the second AND. The following line is apparently similar:

```
IF              (AUTHF AND AUTHM GE 30 AND AUTHS LE 15)ATSHIFT=1
```

However, it would be decoded in a different sequence: GE, LE, first AND, and second AND. If the father's authoritarian score is 20, the mother's is 31, and the student's is 9, the first IF command would not set the attitude shift equal to one, but the second would. Since GE is evaluated first, the second (and incorrect) IF instruction appears as follows:

```
IF              (AUTHF AND (AUTHM GE 30) AND (AUTHS LE 15))ATSHIFT=1
```

Any nonmissing value for the father's authoritarian score (AUTHF) would be accepted.

Inner parentheses are not strictly necessary; the original instruction could be correctly written as follows:

IF   (AUTHF GE 30 AND AUTHM GE 30 AND AUTHS LE 15)ATSHIFT=1

The order of precedence is further confused by the presence of other levels of operation. Table 7.5 gives a complete breakdown of these interrelationships. To demonstrate the complexities, consider the following statement that involves every level of operation.

IF   ((AUTHF AND AUTHM) GE 30 AND AUTHS+(SQ1+SQ2+SQ3+SQ4+SQ5
            +SQ6+SQ7+SQ8+SQ9)/9 LE 20)ATSHIFT2=1

TABLE 7.5. Processing Hierarchy for SPSS (Syntax)

I.   Parentheses
II. Arithmetic operators
     A. Special library functions
         1. SQRT, squart root
         2. LN, natural log
         3. LG10, base 10 log
         4. EXP, *e*
         5. SIN, sine
         6. COS, cosine
         7. ATAN, arctangent
         8. RND, whole number rounding
         9. ABS, absolute value
        10. TRUNC, truncation
        11. MOD10, modulo
     B. Exponentiation, **
     C. Multiplication and division, * and /, respectively
     D. Addition and subtraction, + and −, respectively
III. Relation operators
     A. GE, greater than or equal to
     B. LE, less than or equal to
     C. GT, greater than
     D. LT, less than
     E. EQ, equal to
     F. NE, not equal to
IV. Logical operators
     A. AND, OR, NOT
     B. NOR, NAND, etc., are disallowed

Note: All operators of equal precedence are evaluated sequentially from left to right.

The purpose of such a complex instruction can be seen by recalling that SQ1 through SQ9 were questions on liberalism/conservatism that were scaled from 1 to 9, the low scores indicating a liberal response. By adding the average score to the student's authoritarian score (AUTHS), the researcher might obtain a more cautious assessment of the student's attitudes.

Deeply embedded parentheses are not necessary in the above IF instruction because of the rules of precedence. The order of the above would be as follows: (1) the left-most inner parentheses, (2) the next set of inner parentheses, (3) the division of that inner set by 9, (4) the addition of the result of (3) to AUTHS, (5) GE, (6) LE, and (7) AND. The alternative to careful consideration of the precedence is to force things with many levels of parentheses. Deeply embedded parentheses can be as confusing as the operations' order. Use the style that produces correct logic with the least difficulty.

SPSS provides a means to shorten IF instructions by using implied operators. This often only produces confusion, and even experts experience difficulty interpreting such instructions a year or two after writing them. Beginners should definitely avoid these abbreviations.

The effects of multiple IF commands are cumulative with regard to the SPSS data set. They do not affect each other, provided that each one defines a new variable. IF instructions may be combined by using the result of the first as a part of the second, and so on. Thus, the complexity may be reduced by making each step smaller. Consider the following rewrite of the preceding instruction:

```
COMPUTE      X=0
IF           (AUTHF AND AUTHM GE 30)X=1
COMPUTE      Y=(SQ1+SQ2+SQ3+SQ4+SQ5+SQ6+SQ7+SQ8+SQ9)/9
COMPUTE      Y=Y+AUTHS
IF           (X EQ 1 AND Y GE 20)ATSHIFT2=1
```

This decomposition is not the only one possible. The critical issue remains the same: does this instruction result in a correct manipulation of the data? Careful attention to such instructions will ensure against both faulty analyses and personal aggravation.

The SELECT IF and the *SELECT IF operations function like the IF operation. However, when SELECT IF and *SELECT IF instructions are used in multiples or in combination, they must be joined in certain ways. See Section 7.4.2 for more on this topic.

The SPSS manuals suggest that the user always check the consequences of different values in a complex data manipulation. The reason for this is that the full ramifications are not always apparent; combinations of values occasionally produce logical but unintended results. The solution to this problem is a partial truth table. A truth table is a representation of all possible configurations and whether they are true or false. In much of social science, a complete truth table would be much too large and complex, hence we use the partial table. What is suggested here is the creation of a table with selected values, especially limits and missing values. For example, consider Table 7.6. We can see that when either the mother's or the father's authoritarian score has the missing value (99) and when the student's authoritarian score is less than 20, the statement is true and the attitude-shift value (ATSHIFT) is set at 1. In most cases, the researcher would not consider family authoritarianism (AUTHF + AUTHM) to be determinable when one or both scores are missing. A solution for this case is found by writing another IF instruction that resets ATSHIFT to a value other than 1 when the sum is greater than 78. Thus, if either or both parents' authoritarian scores are missing, the value of the attitude shift does not equal 1.

A truth table for this complex IF instruction would be lengthy but would be of great value in determining the consequences of certain values. For example, if the variables SQ1 through SQ9 all have 0's as the missing values, then their sum and their average will be artificially deflated. A missing value of 0 is neither liberal nor conservative, but it would be considered the ultimate in liberal scores by the IF statement.

All of this concerns us because IF and COMPUTE instructions create new variables that are not limited by the MISSING VALUES instructions. The ASSIGN MISSING command partially solves this problem

TABLE 7.6. *Truth Table for the Instruction* IF ((AUTHF+AUTHM) GE 60 AND AUTHS LE 20)ATSHIFT=1 *(all missing values set to 99)*

Values of AUTHF	Values of AUTHM	+	Sum ≥ 60: T or F?	Values of AUTHS	AUTHS ≤ 20: T or F?	Statement T or F?
0	0	0	F	ALL 10	T	F
0	0	0	F	ALL 29	F	F
0	39	39	F	ALL 19	T	F
21	39	60	T	ALL 11	T	T
21	39	60	T	ALL 38	F	F
30	30	60	T	ALL 07	T	T
30	30	60	T	ALL 20	F	F
99	0	99	T	ALL 04	T	T
0	99	99	T	ALL 19	T	T
99	32	131	T	ALL 09	T	T

by assigning a new missing value to the result of an IF statement when any of the elements is missing. Unfortunately, when the products of a modification result in the assigned missing value, the case will be dropped. Consider the following lines:

```
COMPUTE        FAMAUTH=AUTHF+AUTHM+AUTHS
ASSIGN MISSING FAMAUTH(99)
```

Since the individual missing values are coded 99, it seems natural to also assign 99 to the family authoritarian score (FAMAUTH). However, if authoritarian scores vary from 1 to 39, there are many legitimate combinations that equal 99.

The only overall solution to such problems is the careful consideration of each and every consequence. Logic traps are inevitable if one uses SPSS long enough. If one is skilled enough or if the problem is important enough, traps may even be set for the improper results themselves!

7.4.4   Data Transformations and Conversions

All of the major analysis packages provide powerful procedures for modifying raw data. SPSS is particularly convenient and very comprehensive in this respect. The manuals should be your primary source of information. Only problem areas and special applications are mentioned in this text.

The general form of SPSS transformations can be illustrated with a modification common to psychology: $X = \sqrt{X}$ and $X = \sqrt{X + 1}$.

```
COMPUTE        X=SQRT(X)
COMPUTE        X=SQRT(X+1)
```

In these two examples, many of the basic features of SPSS data modification can be observed.

1. Only one variable modification is allowed per COMPUTE instruction.
2. Multiple operations to the right of the equals sign are allowed.
3. Parentheses are used to establish the order and to note the object of a special function (e.g., SQRT).
4. The effects of COMPUTE commands are cumulative.
5. Only standard arithmetic operations and FORTRAN library functions (SQRT) are accepted.

The first feature does not prevent multiple variables from appearing to the right of the equals sign, as in this example:

```
COMPUTE        SUMFQ=FQ1+FQ2+FQ3+FQ4+FQ5+FQ6+FQ7+FQ8+FQ9
```

It is also obvious in the line above that adding the FQ items in a particular order would be meaningless. SPSS determines the order of operation in COMPUTE instructions using the following rules:

complete operations within parentheses first,
then perform special (library) functions,
then perform arithmetic functions, with exponentiation first, division and multiplication second, and addition and subtraction third,
do the operations from left to right.

Consider this complex transformation:

```
COMPUTE        AVGFAMQ=FQ1+FQ2+FQ3+FQ4+FQ5+FQ6+FQ7+FQ8+FQ9/9+MQ1+MQ2
               +MQ3+MQ4+MQ5+MQ6+MQ7+MQ8+MQ9/9/2
```

And if we wish the results on a logarithmic scale, we have this:

```
COMPUTE        AVGFAMQ=LN10(FQ1+FQ2+FQ3+FQ4+FQ5+FQ6+FQ7+FQ8+FQ9/9)
               +LN10(MQ1+MQ2+MQ3+MQ4+MQ5+MQ6+MQ7+MQ8+MQ9/9/2)
```

The first statement is incorrect if the primary purpose is to add the father's and the mother's answers, to scale them by the number of questions, and then to add the two averages. Because division comes before addition, FQ9 and MQ9 would both be divided by 9 and then added to the other questions, which is not the same as adding each set of questions and then dividing by 9. Try the operation assuming that all the FQ's were answered with numeric 4 and that the MQ's were answered with 5. A correct statement would be this:

```
COMPUTE        AVGFAMQ=((FQ1+FQ2+FQ3+FQ4+FQ5+FQ6+FQ7+FQ8+FQ9)/9
               +(MQ1+MQ2+MQ3+MQ4+MQ5+MQ6+MQ7+MQ8+MQ9)/9)/2
```

or this:

```
COMPUTE        AVGFAMQ=(FQ1+FQ2+FQ3+FQ4+FQ5+FQ6+FQ7+FQ8+FQ9
               +MQ1+MQ2+MQ3+MQ4+MQ5+MQ6+MQ7+MQ8+MQ9)/18
```

There is no confusion with the first COMPUTE instruction because of the precedence rules. In other words, it is unnecessary to include parentheses around the respective summations and divisions. The outer set of parentheses must be used only to prevent the second addition-division combination from being divided by 2.

The attempt to convert the results to a logarithmic scale is of course also incorrect. Arithmetic with logarithms is not the same as arithmetic with standard numbers. However, adding yet another level of parentheses creates a COMPUTE instruction that is very dense and difficult to check. It is less confusing to do complex manipulations in steps:

```
COMPUTE        AVGFQ=(FQ1+FQ2+FQ3+FQ4+FQ5+FQ6+FQ7+FQ8+FQ9)/9
COMPUTE        AVGMQ=(MQ1+MQ2+MQ3+MQ4+MQ5+MQ6+MQ7+MQ8+MQ9)/9
COMPUTE        AVGFAMQ=(AVGFQ+AVGMQ)/2
COMPUTE        AVGFAMQ=LN10(AVGFAMQ)
```

We can also see in this example that the effects are cumulative; i.e., the variable AVGFAMQ is available only as a logarithm and not as an average. The same is true for any modified variable—only the end result is available to SPSS.

To many readers, the problem of precedence is no doubt familiar from the ambiguities of standard algebraic notation. Lack of ambiguity is why some calculators use reverse polish notation (RPN). The best advice is to use parentheses whenever you are in doubt as to the result. It is also advisable to work a few test examples with some of the actual data.

Another useful application of SPSS data modifications is in the calculation of time between dates. The convention of writing day/month/ year is of little computational use and is potentially misleading. When a birth date (23/03/60) is subtracted from an interview date (17/11/81), it does not produce the subject's age. It helps to reverse the dates to year/month/day, but this still lacks the numeric properties expected of time measurements. Consecutive days from one date to the next is the most desirable way to measure. In SPSS Version 6, this must be calculated, but in Versions 7, 8, and 9, there is a special function named YRMODA that can be used (see Section 7.4.5). As an example for users of Version 6, consider a survey of college freshmen in which the age in days is desired. To achieve this, we must use another powerful SPSS transformation, the IF instruction. Its general form is as follows: if X is true, then Y equals some arithmetic expression. For example,

```
IF          (AGE GT 70)RETIRED=1
```

For the moment, it is sufficient to know that this statement is translated as: if the variable AGE is greater than 70, then the variable RETIRED equals 1. Assuming that the dates in question are all in the year/ month/day order, we could determine the birth dates and the testing dates as follows:

```
IF          (YEAR EQ 1960)YEARDAYS=0
IF          (YEAR EQ 1961)YEARDAYS=365
IF          (YEAR EQ 1962)YEARDAYS=730
IF          (YEAR EQ 1963)YEARDAYS=1095
IF          (YEAR EQ 1964)YEARDAYS=1460
```

The same type of procedure for months and days gives birth date in terms of days from an arbitrary starting point—in this case, January 1,

1960. By adding the three (COMPUTE BDAYS=YEARDAYS+MONTH-DAYS+DAYS), we complete the calculation.

In determining the number of consecutive days to the interview date, year calculations are not necessary if all subjects were interviewed in the same calendar year. Consider in this example that the interview year is 1981, the month is November, and that all subjects were interviewed on days 1 through 6.

```
COMPUTE        TESTYEAR=7300
COMPUTE        TESTMNTH=304
COMPUTE        TESTDAYS=TESTYEAR+TESTMNTH+TESTDAYS
COMPUTE        AGE=TESTDAYS-BDAYS
```

In this way, each case can have its age determined to the day. This assumes that there are no freshmen under 16 or over 21 and that all interviews are done in one month. By generalizing this procedure, however, any age is calculable. Leap years are ignored in this procedure.

Two additional points about SPSS data modifications are illustrated by this age-calculation example. The most obvious point is the way many different modification instructions may be combined in almost limitless variations. The only trick is to maintain the logic correctly and to avoid creating impossible syntax. It may seem elementary, but a very common problem even for experienced SPSS users is logic that does not result in the desired goal. The second additional point is the use of the COMPUTE instruction to define values. TESTYEAR=7300 is not ordinarily thought of as a computation. Nonetheless, it sets a value and is thus an elementary COMPUTE.

7.4.5   Special Features of Version 9

Version 9 of SPSS includes a number of pro technique features that are beyond the beginning level.

YRMODA is extremely useful for calculating the number of days elapsed between any two dates. The following whimsical example determines George Washington's age if he had lived until July 4, 1982.

```
COMPUTE        BYEAR=1732
COMPUTE        BMONTH=02
COMPUTE        BDAY=22
COMPUTE        DYEAR=1982
COMPUTE        DMONTH=07
COMPUTE        DDAY=04
COMPUTE        AGEDAYS=YRMODA(DYEAR,DMONTH,DDAY)-YRMODA(BYEAR,BMONTH,BDAY)
```

All of this could be shortened to the following:

```
COMPUTE        AGEDAYS=YRMODA(1982,07,04)-(1732,02,22)
```

To obtain Washington's age in years, we would divide by 365.25. Any dates between October 15, 1582 (which is day one) and the slightly ridiculous date of December 31, 47,516 (for IBM computers only) may be obtained. Because dates are converted to continuous days, a great many calculations are possible that maintain the metric properties we expect of time analyses. An additional desirable feature is the automatic treatment of leap years.

The interaction of YRMODA with other data-selection instructions can be quite complex. See Section 11.11 of *SPSS UPDATE, 7–9*, page 278.

Another useful feature found only in Version 9 is the ability to execute SPSS analyses without knowing the number of cases. Indeed, the N OF CASES instruction may be omitted under most circumstances. When this is done, UNKNOWN is assumed, and an END INPUT DATA instruction is required at the end of the raw data. When each case requires several terminal lines or cards, the raw data should be followed by an END INPUT DATA instruction for every additional line of data per case.

A potentially confusing issue for those who have used SPSS for a long time is the phrase "comand file." A command file is the same as a control card deck in that both contain the instructions for SPSS to follow. Typically, a command file is created at a terminal and does not contain data.

The authors of SPSS have also extended its format facilities to include B, C, R, and Z formats. This applies to IBM machines only. Careful checking with those organizations that prepare SPSS for use on DEC (Digital Equipment Corp.) or other machines is necessary to determine which formatting options exist for the machine you are using. Your local consultant may or may not know the specifics.

There are a fair number of modifications and extensions in Version 9. SPSS users should be sure to check the update manuals when using any of the following keywords: EQ, DO REPEAT, SORT CASES, and SAMPLE.

A final factor is the modification of many of the calculation procedures to what is known in FORTRAN as double precision. That is, twice the number of significant digits are used in computations. This gives greater precision and eliminates some of the oddities that occur due to rounding. Double precision can be attained only in certain combinations of manufacturers, machine models, and operating systems.

7.5   Pro Tips and Techniques for Specific Subprograms

There is no single correct way to conduct an SPSS analysis. Even "pros" will not all use SPSS in the same way. This section is a summary of this author's experience and is not the final word on analysis. The comments that follow should be read keeping in mind that the techniques will require further adaptation to individual needs. Also, this is not a substitute for reading the SPSS manual itself.

7.5.1   The CONDESCRIPTIVE and FREQUENCIES Subprograms

It is often desirable when performing an analysis to obtain basic descriptive information on all variables before going on to do a more elaborate analysis. The CONDESCRIPTIVE subprogram is ideal for this purpose. It is also the type of run that is normally used to establish an SPSS system file for later use. (Caution should be exercised when requesting the various statistical options, since SPSS cannot determine the legitimacy of a given procedure. For example, while skewness would be inappropriate for ordinal data, median, mode, and range are perfectly in order.)

Particularly useful is the calculation of Z-scores and the placing of these scores instead of the original scores in an output file to be used in further analyses. Exercise care when the SAMPLE or the SELECT IF procedures are used with Z-scores because they interact. Similarly, thought must be given to when other data modifications should be executed, i.e., before or after calculating the Z score.

The CONDESCRIPTIVE subprogram may be used as a general data clean-up procedure. Each variable may be examined for appropriate values. It is very useful in large data sets to initially define not only the variables but also the gaps between the variables (if any). For example, if ID and AGE were the first two variables in a set of 10,000 cases and if there were a blank between each, then the definition of the space as a variable will allow CONDESCRIPTIVE to examine that space. Obviously, nothing but blanks should be discovered. While it is true that the spaces are effectively ignored in both fixed and free formats, the discovery of data in one of these locations often means that one of the variables on either side of the space has been inadvertently moved into the blank. The effect of such data-preparation errors is different for fixed and for free formats. However, both would be in error.

The CONDESCRIPTIVE subprogram effectively checks the structure of the data. Generally, it is unwise to create an SPSS file from the checking procedure suggested here. A final note on the CONDESCRIPTIVE subprogram: the STATISTICS default is ALL!

The FREQUENCIES subprogram plays much the same role as does CONDESCRIPTIVE, except that it is for discrete variables. Here we immediately see a need for caution. Calculating frequencies on continuous variables can cause embarrassing and costly problems. Grade point average, most often a variable coded as a three-digit number with two decimal places, has a range from 0.00 to 4.00 (or to 5.00). SPSS would produce a table of frequencies 400 or 500 lines long (at 20 rows per page times 20 pages). Continuous variables can be held in check by first recoding the variable. The number of discrete categories used is naturally the user's decision.

Another area of some difficulty in the FREQUENCIES subprogram is the GENERAL versus the INTEGER mode. The GENERAL mode will obtain frequencies for any data type. The INTEGER mode excludes decimal and alphabetic data. When data sets are small, the difference is minimal; but with large data sets, INTEGER saves time and memory allocation, which often translates into dollars saved. The solution is to recode decimal and alphabetic data. Users occasionally object to the INTEGER mode on the grounds that the minimum and the maximum scores are unknown. There are two answers to this objection. First, while the actual minimums and maximums are often unknown, the theoretical limits are not. Thus, if an attitude questionnaire contains 39 true-or-false items, the minimum and the maximum scores would be 0 and 39, respectively. If would be of little immediate interest that the actual range was from 4 to 35. Second, option 9 in the CONDESCRIPTIVE subprogram provides the necessary information, that is, the range of a variable. This is another advantage of running CONDESCRIPTIVE first.

Several of the options available with the FREQUENCIES subprogram merit special attention. Option 3 restricts the printing to $8\frac{1}{2}$ by 11 inches, so the output can be included in manuscripts. Option 3 is also useful when working from a terminal, since it helps to avoid the broken lines that result when the SPSS package types a long line on a short screen. (This is not needed in SCSS, which is already oriented toward terminals.) Option 5 produces a condensed printing format, which is quite helpful when working with large data sets. Histograms are available through option 8. Be cautious, though, as you might not want a graph of every variable. Define a second task with only the variable of interest.

A final word about the FREQUENCIES subprogram: the STATISTICS default is NONE—the exact opposite of the CONDESCRIPTIVE subprogram's default of ALL.

7.5.2   The CROSSTABS Subprogram

The CROSSTABS subprogram is very similar in form and in structure to the FREQUENCIES procedure. It, too, has both INTEGER and GENERAL modes. The same advice holds: use the INTEGER mode whenever possible. In addition to the advantages of speed and storage, more statistics are available in the INTEGER mode.

One peculiarity of this subprogram is the limits on the use of implied lists, e.g., ITEM 1 TO ITEM 17. This list form may not be used with the INTEGER mode. As with most SPSS programs, the real problem is to restrict the output to exactly what is required. This is more than a casual admonition. CROSSTABS can produce complicated tables with innumerable subtables (for example, a table of sex by age by college class by major by SAT score by GPA, etc.). Not only are the resulting tables massive, but they are almost impossible to interpret. It is essential that the actual needs are carefully examined in this procedure.

In the CROSSTABS subprogram, be sure to note that the INTEGER mode requires both a VARIABLES= and a TABLES= instruction, whereas the GENERAL mode requires only a TABLES= instruction.

7.5.3   The AGGREGATE Subprogram

The AGGREGATE subprogram is a highly specialized technique. It obviously produces statistics in aggregate form, thereby allowing phenomena to be viewed from many different perspectives. The researcher interested in health care delivery might collect raw data at the level of the individual recipient, but there are several other levels of interest, including statewide, regional, and even larger political areas. AGGREGATE provides a structured means of obtaining these interrelated but divergent views. Using AGGREGATE presents several points of concern:

1. Cases must be previously sorted on the grouping (aggregate) variables; e.g., cases must be adjacent.
2. Similarly, if cases are aggregated on four variables (the maximum) and if one or more of the variables is missing, then the entire aggregate is excluded. This is a variation on the classic empty cells problem.
3. All variables must be numeric. This does not mean that grouping variables need be numeric—only those with which actual statistics are performed.
4. Data modification and sampling procedures markedly affect the results of the AGGREGATE subprogram, so be cautious.
5. Option 3 is very useful in that it maintains the original identifiers for the grouping variables. It is difficult to remember just how an analysis was performed after a year or more has passed. When terminals are used, one may not even have a copy of the original program instructions to jog one's memory.

6. The AGGREGATE subprogram uses binary output, which provides a processing advantage when working within SPSS. However, binary files are so unique that they are compatible within only one combination of computer, analysis package, and computer center.

7.5.4   The BREAKDOWN, CROSSBREAK, and T-TEST Subprograms

The BREAKDOWN subprogram is similar to the CROSSTABS and the FREQUENCIES subprograms. The distinction between the INTEGER mode and the GENERAL mode is made with the same attendant advantages and disadvantages. Remember that the INTEGER mode requires that one specify the VARIABLES= and the TABLES= instructions, while the GENERAL mode requires only the TABLES= instructions.

Users of the BREAKDOWN subprogram normally experience problems with zero cells and with severe case deletion due to missing values. Generally, the restriction that is more troublesome is that the dependent, or criterion, measure must be numeric. The variable named first is automatically considered to be the dependent measure. Obviously, violations of the first restriction will result in an error condition. However, to SPSS all variables are potential dependent measures; thus, the error cannot be detected. The STATISTICS option is also troublesome. It applies only to one-way breakdowns. Although there are statistics for multiway situations, they are not available in SPSS.

The CROSSBREAK subprogram usually presents few problems, but you should be selective in choosing statistics with this procedure. Even though the statistics are generally appropriate, they tend to be somewhat esoteric.

The T-TEST subprogram, despite its apparent simplicity, has a number of areas where users commonly make mistakes. Confusion arises over the exact type of t-tests available. They include independent samples, paired samples (correlated), and the t approximation for populations with different variances. All tests are two-tailed. The GROUPS= and PAIRS= instructions request the first two types, respectively. The unequal variances option is determined by the F statistic on the sample variances. If the sample variances are significantly different, you will need to use the nonpooled estimate. When specifying GROUPS=, use caution because there are several distinct ways to accomplish this end. Defining group 1 as all cases greater than or equal to some value is not the same as defining group 1 as all cases equal to some value. Careful reading of the SPSS manual's description of these options will reveal the differences. In short, groups may be defined in a variety of nonequivalent ways in this subprogram. The PAIRS instruction also produces certain problems. SPSS cannot make a judgment about the advisability of a specific paired t-test. All variables in a study could be compared with each other. This would result in an overly large number of tests, many of which would be pure nonsense, such as those involving ID number and grade point average. As in all analyses, the user should choose carefully and should not allow the system to perform indiscriminate tests.

SPSS is also insensitive to errors in the scale of measurement. The T-TEST procedure seems particularly prone to tests that violate statistical assumptions. While a GPA is an interval measurement, an ID is strictly nominal. In paired t-tests, both variables must be interval measurements. The fact that SPSS can execute the statistics does not imply the correctness of the numbers, much to the disappointment of many students.

7.5.5   The DISCRIMINANT, FACTOR ANALYSIS, GUTTMAN SCALE, and CANCORR (Canonical Correlation) Subprograms

If you are ready for the DISCRIMINANT, FACTOR ANALYSIS, GUTTMAN SCALE, and CANCORR (canonical correlation) subprograms, you are probably an expert and do not need the little advice that could be given here beyond what the SPSS manual offers. On the other hand, if you need and understand these techniques but are unfamiliar with the SPSS package, it is strongly suggested that you practice by obtaining several preliminary analyses. Do some crosstabulations or some frequency analyses, and obtain a few simple correlations. In this way, you will become familiar with SPSS before attempting these subprograms. These procedures are simply not a good place to begin because they would probably intimidate potential users.

7.5.6   The REGRESSION Subprogram

The description of the REGRESSION subprogram in the SPSS manual constitutes a minor treatise on the general topic of regression analysis. A careful reading is recommended for all users. If you are new to SPSS or to regression, avoid the complications attendant upon inputting a matrix of correlations or means and standard deviations. The best advice is to practice producing the many different types of regression with a simple data set.

The OPTION and STATISTICS instructions are critical for smooth analyses, and they should be given careful attention. Extra attention should also be given to the treatment of missing values and cases. Unless you are a regression expert, be wary of specifying large numbers of variables and complex subfile structures. As with all complex procedures, until you know what you want and how to get it, resist the temptation to allow SPSS to produce everything that it can.

The cautions listed above are not meant to frighten potential users of REGRESSION. A carefully planned approach works best. By all means, practice with data that are not of personal interest—it will help you to concentrate on the technique rather than on the results.

7.5.7   The PARTIAL CORR (Partial Correlation) Subprogram

In many research efforts, the partial correlation subprogram (PARTIAL CORR) is a very useful and powerful technique. Those students who are just starting to learn this procedure seldom experience difficulty with specific areas. Instead, they have trouble learning enough about the requirements to begin practice. One can know what a partial correlation is and still have difficulty with this subprogram. Experience indicates a solution. Do not begin with a matrix of intercorrelations (matrix input); rather, use paired observations. In this sense, the PARTIAL CORR subprogram functions just as the PEARSON CORR and NON-PAR CORR procedures (see Section 7.5.8). Choose the variable to be correlated and then the variables that are thought to affect the variables of interest (control variables). Finally, request the appropriate orders of partial correlations. For example, correlating GPA with a conservatism/ liberalism score, controlling for effects of age, sex, and race, and desiring first- and third-order partial correlations, we would have the following:

```
PARTIAL CORR    GPA WITH CONLIB BY AGE,SEX,RACE(1,3)
STATISTICS      1
```

This will produce the correlation between GPA and conservatism/liberalism (CONLIB) (zero-order partial) and three first-order partials: GPA with CONLIB controlling for AGE; GPA with CONLIB controlling for SEX; and GPA with CONLIB controlling for RACE. One third-order partial GPA with CONLIB controlling for AGE, SEX, and RACE will also be computed.

Thus, a little practice with PARTIAL CORR is very useful for the first-time user. Several points seem to remain difficult despite practice. First, it is impossible to specify a fifth-order partial when only three controlling variables have been designated. Second, the limit on partials is not five; rather, only five partial values may be selected. With ten control variables, one could choose (1,3,5,7,9), (2,4,6,8,10), or (6,7,8,9,10), but not (1,2,3,4,5,6,7,8,9,10). Last, the standard (or zero-order) correlation is produced only when 1 is specified in the STATISTICS instruction.

7.5.8 The PEARSON CORR (Pearson Correlation), NONPAR CORR (Nonparametric Correlation), and SCATTERGRAM Subprograms

The Pearson correlation (PEARSON CORR) subprogram is very clear and direct in function. Experience indicates only two minor problem areas. First, when intercorrelation matrices are desired, the WITH instruction is not used. By choosing the correct option, we may select any matrix to be output for further use. However, regular paired variable correlations may not be selected. The second problem is the tendency for the researcher to allow SPSS to correlate everything. Doing this with the PEARSON CORR subprogram often results in a small mountain of coefficients.

The nonparametric correlation subprogram (NONPAR CORR) is nearly identical to PEARSON CORR except that the correlations are the Spearman or the Kendall variations. The only tricky area is in selecting the type of statistics desired. These types are documented in the manuals, but they are not clearly shown. Briefly, the instruction OPTION 5 produces Kendall correlations; the instruction OPTION 6 produces both Kendall and Spearman correlations, and specifying no option produces only Spearman correlations.

The SCATTERGRAM subprogram is also very direct and warrants only brief comments. First, the instruction PAGESIZE does not limit the size of the printed plots, which will be of fixed size. Second, easy readability and interpretation are achieved by advanced planning of the plots. Third, unlike previous recommendations, ALL is recommended as the STATISTICS option. Performance is only slightly affected, and the statistics are almost always of some interest.

7.5.9 The NPAR TESTS (Nonparametric Tests) Subprogram

The nonparametric tests subprogram (NPAR TESTS) is among the easiest and most direct of all SPSS procedures. Problems—when they occur—are usually errors of judgment, e.g., using an inappropriate test or supplying insufficient information to complete a specific test. Such problems apparently occur because certain users assume that since nonparametric tests are assumption-free, then anything is allowable. This is simply not the case. The user should carefully match the assumptions and functions of each test to the data. Further, searching among tests in an attempt to find one test that indicates significance is, in a word, dishonest.

7.5.10 The REPORT, RELIABILITY, and SURVIVAL Subprograms

The three subprograms REPORT, RELIABILITY, and SURVIVAL are relatively new, so little experience with them has been accumulated.

Clearly, RELIABILITY and SURVIVAL are very useful procedures, but they are hardly topics for beginners and are certainly not suitable for learning basic SPSS operations.

The SURVIVAL subprogram will prove of interest not only to those researchers who traditionally work with life statistics but also to ethnologists, ecologists, and comparative psychologists. RELIABILITY is of most interest to those who work in test development and theory. Two interesting aspects of the RELIABILITY subprogram are its ability to perform limited-item analyses and repeated-measures ANOVA (analysis of variance). The REPORT subprogram does precisely what its name implies—it writes reports in a variety of forms. This should prove quite useful to those researchers who must provide descriptive information on large data sets. Typically, this material would be used as an appendix to a dissertation or other project. The point is that REPORT allows greater tailoring of the output than is provided by typical SPSS programs.

7.5.11   The ANOVA and ONEWAY Subprograms

One of the more common problems with the ANOVA (analysis of variance) subprogram is defining what it will and will not do. SPSS ANOVA is *not* any of the following:

a multivariate program in the sense of multiple dependent variables,
a repeated measures program,
a nested, or fractional replication, program,
a random-effects program,
a mixed-effects program,
an unweighted means analysis,
a program that is able to assess interaction between variates and
    covariates,
a least-squares solution.

Despite this list of exclusions, ANOVA is a powerful procedure that can analyze a fixed-effects model from the classic experimental approach, the hierarchical approach, or the regression approach. Additionally, covariates may be included and treated in a variety of ways. Variate-covariate interaction must be assessed through the REGRESSION subprogram or other techniques. Finally, multiple classification analyses (MCA) may be selected.

Versions 7 and 8 contain a repeated-measures feature. However, it is obtained through the RELIABILITY subprogram, which places several constraints on it. In effect, the repeated-measures feature is a scale composed of multiple items, where each item is a repeated measure. Thus, this addition is restricted to one-factor designs—the repeated factor. See the SPSS manual for a discussion of RELIABILITY.

The instructions necessary to execute subprogram ANOVA are typical of SPSS. There are several points of occasional confusion, however. First, the minimums and the maximums for each independent variable must be specified (recall this requirement with the FREQUENCIES subprogram). It is allowable to define limits sufficiently large to cover all independent variables, but this results in memory allocations that are unnecessarily large. Second, more than one dependent measure may be specified, but this gives us independent analyses. Third, although the usual SPSS conventions for implying lists of variables are valid, caution should be exercised here because many unnecessary and unwanted analyses can be produced. Fourth, as many as five covariates may be included with each analysis.

Many people are confused about the use of the terms metric and nonmetric to describe variates and covariates, respectively. The terms refer to the typical situation where the independent variables are factorial classifications, such as treatments A, B, and C, and where covariates are continuous variables, such as age and IQ. This need not be the case, however. If treatments A, B, and C represent the drug dosages 500 mg, 1000 mg, and 1500 mg, then they would be three points in a continuous range of dosages. The reverse may occur for covariates. For example, there may be six different sedatives routinely given to psychiatric patients. Such medications might not form a continuous metric scale of sedation, but the researcher might desire to remove them as sources of variation in a given ANOVA. This discussion does not eliminate the restrictions on the measurement of dependent variables; we must have at least interval measurements.

The ONEWAY subprogram is not fundamentally different from a one-way analysis with ANOVA. However, the number of additional features available makes ONEWAY highly desirable where applicable. Most notably, it is used for trends, for *a priori* and *a posteriori* contrasts, for fixed- and random-effects models, for homogeneity tests, and for pooling. The only specific advice is to be wary of complex analyses until you have gained a little experience with this procedure.

7.5.12   The MANOVA (Multivariate Analysis of Variance) Subprogram

With the addition of MANOVA (multivariate analysis of variance) in Version 9, SPSS has made an important advance. Many experimentally oriented researchers did not use SPSS in the past because the ANOVA statistics did not meet their needs. MANOVA offers a large number of design options for these researchers. The procedure can handle the usual designs and those with unequal n's, with repeated measures, and with incomplete or confounded designs. In addition, MANOVA can accommodate multivariate tests, covariance, and multivariate multiple linear regression, making this subprogram a powerful tool indeed.

The beginner should practice with the simple designs first. It is also recommended that all who are new to this procedure carefully study the section on MANOVA commands. Some caution should be exercised because, as with all new techniques, there are likely to be a number of hidden bugs. It is a good idea to check the procedure against known results. The more complex a procedure is, the more likely it is that errors will be found. No software developer could afford to check all of the possibilities.

MANOVA offers several techniques (SSTYPE) for unequal cell sizes, but these methods are explained only in the section on MANOVA commands. A final comment—the cell information (CELLINFO) option offers a quick check on the correctness of the analysis by providing information on cell means, sums of squares, and so forth.

7.5.13   The BOX JENKINS Subprogram

The BOX JENKINS time-series analysis is a fairly advanced statistical technique and as such is not a good place to begin to study the SPSS package. However, if you need to forecast the future based upon seasonal or other calendar-related events, this is a useful method.

7.5.14   The NEW REGRESSION Subprogram

The NEW REGRESSION subprogram covers the same statistical area as the REGRESSION procedure as well as a good deal more. Both functions are maintained in Version 9 because of their significant differences in syntax. Those who are familiar with the older procedure will need a bit of study before attempting an analysis with the newer one. Beginning researchers should not attempt to learn SPSS with the newer procedure. Indeed, regression as a statistical topic is probably easier to learn with the older subprogram. The real virtues of NEW REGRESSION are its impressive flexibility and its up-to-date techniques. The result is a complex set of instructions. If a researcher understands regression and needs this versatility, then NEW REGRESSION is excellent. There is one caution, though: within each choice of regression method will be several choices regarding the type of measurement preferred.

7.5.15   The GRAPHICS Subsystem

At the moment, the GRAPHICS subsystem of SPSS is an extra-cost option. With this option, you can generate impressive-looking charts and graphics by using fairly standard SPSS instructions. The information must be generated by SPSS and then passed along to other processors that in turn control the physical devices (such as plotters) that produce the charts and graphs. The real advantage is that the user need not learn to instruct a graphics printer or plotter.

8

SAS:
Versions 79 and 79.5

For anyone making their first attempt to learn SAS (Statistical Analysis System), there are three basic problem areas, the first of which is input. While input is an important component of all computer system packages, it assumes monumental proportions in SAS. There are input options, variations on options, and variations on variations. The rest of SAS is, however, remarkably straightforward if one has a good basic knowledge of statistics, the second problem area. If you know your statistics, then SAS is a smooth and efficient package. The third factor that can cause trouble is the principal language in which SAS is written —PL/1 (Programming Language One). There is nothing wrong in using this language, but it often leads to confusion for those who have been exposed to FORTRAN. (Indeed, the formatting principles in Chapter 6 are oriented to FORTRAN.) Since SAS runs only on IBM machines, the SAS manual is sprinkled with both PL/1 and IBM terminology.

Accordingly, a major portion of this chapter will be devoted to SAS input and file manipulation. The remainder will consist of an applied section for practice with SAS and a "Pro Tips and Techniques" section. The place to begin, though, is with a general overview of the SAS process, which is the purpose of the first section.

8.1 Basic Organization of SAS

There are three basic organizational principles in SAS: the DATA step, the procedure step (PROC), and the data sets (files) and their manipulation. Although other analysis systems have input procedures (the DATA step), select statistics (the PROC step), and utilize files, none are as explicitly organized around the manipulation of data sets as is SAS. What this organization of data sets means is that SAS routes everything to a file. There are data sets of the moment, data sets of the past, and new data sets to create. Indeed, some SAS procedures cannot produce a single line of print. Instead, the user must use one procedure to produce a data set and then use a different procedure to print the results. Some SAS procedures demand certain arrangements before they can be correctly used, thereby requiring still more manipulations.

Consequently, with SAS there is less emphasis on the actual procedures. Many are extremely straightforward and possess only basic options. Most of the flexibility of SAS is found in the DATA step. A typical analysis involves the following actions:

inputting and establishing a new data set or retrieving a preestablished set,
manipulating the data set (or sets) for proper analysis,
requesting the analysis and its options, and
producing output (if this is not an automatic part of the procedure step).

All SAS functions are organized around the concept of the data set. For the social scientist, each set consists of measurements in a particular area of research. The term variable is used exactly as in other program packages; it refers to weight, age, sex, test scores, question responses, etc. However, each individual (rat, person, factory) is considered to be an *observation,* whereas BMDP and SPSS packages refer to cases or to respondents instead. This is a matter of linguistic choice and is not an organizational difference. The same system of variables in columns and observations (cases) in rows is maintained in SAS. Thus, a set of records intended for SPSS is also properly organized for SAS. Further, the term multiple observations refers to multiple cases or to multiple subjects and must not be confused with the term repeated (multiple) measurements. The daily weight of a person in a weight-reduction experiment is an example of one observation with repeated measurements. In short, SAS uses the same structure as the other systems but defines certain words differently.

There are several stylistic aspects of SAS that should be noted:

1. All SAS statements (lines or cards) end with a semicolon (;).

2. Blanks are used as delimiters (separators), such as in this statement: DATA MEN WOMEN. Commas are not used as separators.
3. Statements may begin and end anywhere on a line or card, with continuation from one line or card to the next.
4. The word FORMAT indicates a special type of input that is not typical.
5. SAS does not utilize the input/output aspects of FORTRAN. Rather, PL/1 is used.
6. SAS may be run from a terminal in two ways: it can be run as a batch execution, just as if the terminal lines were cards; or it can be run interactively, where the system conducts a dialogue with the user about the analysis to be done.
7. Abbreviations are restricted and are therefore not recommended.
8. When the equals sign is used, blanks are not required but are allowed, e.g., DATA=PRACTICE or DATA = PRACTICE.

SAS variable names are restricted by the following rules:

1. The names must be eight characters or less in length.
2. The first character must be a letter (as in TESTONE); the remaining characters may be letters or numbers (as in TEST2).
3. Special characters are not allowed (see the exception in item 6 of this list).
4. Blanks are considered to be special characters.
5. De facto blanks may be created using the underscore, for example, TEST_TWO.
6. When a name must include a special character, the name must be enclosed in single quotes (for example, 'TEST/TWO'). (*Note:* SAS uses the term single quote for what is the apostrophe in the other packages and in English usage.) When a name already contains a single quote, substitute two consecutive single quotes (as in 'JIM''S ID').
7. The use of SAS instructions (called *keywords* by SAS) as variable names should be avoided. While this practice is not automatically the cause of problems, it is valuable to know whether an error has been caused by you or by SAS.
8. The general form _XXXX_ should be avoided because SAS generates a number of automatic variables with this form.

8.1.1 SAS Data Sets

All SAS data sets begin with a statement in one of the following formats (with the first format a name is assigned by SAS):

```
DATA;

DATA _NULL_;

DATA PRACTICE;

DATA PRACTICE MORPHINE BEER;
```

where the words PRACTICE, MORPHINE, and BEER are data set names. The above statements would produce one, one, one, and three data sets, respectively.

With terminals or cards, the next statement is the INPUT statement, which is the logical equivalent of the formatting procedure in the other packages. Next comes the CARDS statement, followed by the data deck itself. The final statement contains the instructions for the actions to be taken. These are equivalent to the PCL (program control language) of other packages.

With a user-created tape or disk, the order of the statements is as follows:

```
DATA PRACTICE;
INFILE . . . ;
INPUT . . . ;
(insert further PCL instructions here)
```

When additional data sets are needed for an analysis, the SET instruction is employed, as follows:

```
DATA NEWPRACT;
SET PRACTICE;
(insert further PCL instructions here)
```

In this simple case, NEWPRACT would contain exactly the same information as PRACTICE. This appears to be an unnecessary complication, but it is in fact clear when the nature of SAS data set operations is explained. The essence lies in the concept of multiple files with only one current file.

Figure 8.1 gives a diagrammatic presentation of SAS file structure. Observe from this figure that the DATA instruction always—with one specific exception (see the discussion below on DATA_NULL_)—creates a current working data set. This means that if there are six data steps, the last set created is the current one. The first five would be temporary or library sets, depending on how they were named. Data set names are in reality double names. In the preceding example where the data set is referred to as NEWPRACT, its whole name is WORK.NEWPRACT. The qualifier WORK indicates a working, or temporary, file and is assumed to be the first part of the name unless something else is specified.

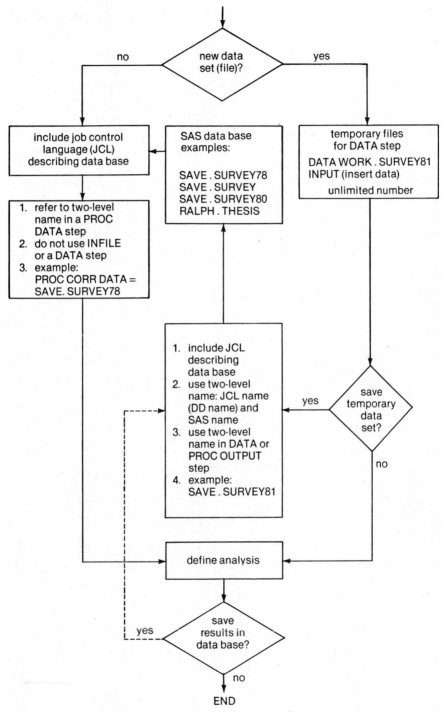

FIGURE 8.1. SAS File Structure and Access Routes

193

194

The user of SAS has the option of permanently storing the current file (see Figure 8.1), in which case the first part of the name is SAVE (for example, SAVE.SURVEY81). When a data set has been saved, its full name must be used; with temporary sets, this is not necessary. When a series of data sets is saved, only the most recent one is considered current. Temporary and save instructions may also be mixed in any order. The essential point is that analyses are performed on the current file only. For example, with the following instructions (the instruction PROC ANOVA calls for the analysis of variance procedure), only EXP2 would be analyzed:

```
DATA EXP1;
INPUT . . . ;
CARDS;
(insert data here)
DATA EXP2;
INPUT . . . ;
CARDS;
(insert data here)
PROC ANOVA;
```

There are two ways to avoid this with SAS. One can include a separate PROC ANOVA instruction after the first data instructions, or one can write a second PROC ANOVA instruction following the one already there and naming the data set to be analyzed, as follows:

```
PROC ANOVA;
PROC ANOVA DATA=EXP1;
```

It is not necessary to state WORK.EXP1 because this is the assumption in both the procedure and the data steps. However, if EXP1 had been saved, its name would be SAVE.EXP1. If no name is given in the data step, SAS will assign the name WORK.DATA1 and then WORK.DATA2, etc. Such names may be used, but the risk of confusion is an obvious disadvantage.

The DATA_NULL_ statement creates a special SAS data set—one that is not intended for analysis by a particular procedure step. This statement avoids creating an SAS data set, but all of the processing (including output) that is possible with a data step may be performed.

8.1.2 Summary

The previous discussion only describes the surface of the SAS data step. Data steps can be merged, sorted, modified, updated, and altered in a variety of ways. The beginner should know that these variations exist but that they are by no means required. Overall, SAS demands that the user name the data set, define its characteristics, modify the set if necessary, and pass it on to the procedure (analysis) step. The details of SAS input are the next logical topic, and the next section is directed to that end.

8.2   Specifics of SAS Input

The classic FORTRAN style of format instruction was discussed at some length in Chapter 6. SAS uses a very different but closely equivalent system (PL/1) of formatting, which is not surprisingly called formatted input. There are three basic forms of formatted input. Numeric input has the general form of $w.$, decimal input has the general form of $w.d$, and character input has the general form of $\$w$. Recall that in these systems of notation, w is the width of the variable and d is the number of decimal places. The dollar sign ($) denotes a nonnumeric value whose width is w. What SAS does not do, however, is use separate variable lists and format instructions. These are combined in common usage. Note that there is a specific exception to this, but it is a minor variation and is discussed in Section 8.6.3. SAS also does not use format symbols, such as I, F, A, H, and X, and it does not use the word FORMAT.

The instruction (keyword) INPUT is used to signal SAS. Typical formats are as follows:

```
INPUT IDN 4. AGE 2. NAME $20 GPA 3.2;

INPUT TRIAL1 4.1 TRIAL2 4.1 TRIAL3 4.1 TRIAL4 4.1;

INPUT NAME $15 Q1 $1 Q2 $2 Q3 $1 Q4 5.;
```

When the format is repetitive, a multiplier may be used to simplify it. The second line above would thus become

```
INPUT TRIAL (4*4.1);
```

However, there is no individual naming of the trials. To accomplish this, SAS uses the implied list, as in

```
INPUT (TRIAL1-TRIAL4)(4.1);
```

The only restriction is that nesting (parentheses within parentheses) is not allowed. Mixing is allowed, though, as in the following:

```
INPUT (TRIAL1-TRIAL4)(4.1) TRIAL5 5.1;
```

Some psychological assessment tests force the subject to choose either A or B; in this case, this might be the format:

```
INPUT IDN 2. (Q1-Q50)($1.);
```

Here IDN would be considered numeric input, and Q1 through Q50 would be considered character input (A's and B's). Note also that the name given to the identification variable is IDN in these examples. This is because ID is an SAS keyword and should not be used as a variable name.

All of the preceding should be reasonably clear to those who are familiar with FORTRAN or who have read Chapter 6. Note, however, that all of the examples given presume data that are contiguous (have no spaces). SAS does not use X- or T-format instructions for skipping. It uses a "pointer" instead. The SAS pointer is a unique and powerful entry tool that is somewhat similar to the cursor on a terminal. The function of the pointer is the same as that of a locator; i.e., it indicates a line and a column. Internally, this is the system's means of maintaining order, and the user has control of the pointer. It can be used for column skipping, and it also allows backtracking within columns and lines. In effect, the user can reorganize the data at the moment of entry.

To demonstrate the power of the SAS pointer, let us consider a survey intended to determine the size, nature, and location of farms in the United States. For such a survey, we might have this format:

```
INPUT IDN $4. STATE $2. COUNTY $2. TOWNSHIP 3. SECTION 3. ACRES 4.1 TYPES $1.
      MAINCROP 2. SECONDCR 2. THIRDCR 2.;
```

Written this way, each observation would be a continuous block of numbers and letters; terminal lines or card columns 1 through 25 would be occupied. As emphasized throughout this text, such blocks are difficult to proofread, and one is likely to commit errors when creating them. If spaces were inserted, FORTRAN would use the X convention. With SAS, the pointer is used. The each sign (@) followed by a number serves this purpose. (*Note:* Do not read the sign @ as the word *each* because it will cause confusion.) Look at this example:

```
INPUT @10 IDN $4 @20 STATE $2 COUNTY $2 TOWNSHIP 3. SECTION 3. @35 ACRES 4.1
      @40 TYPE $1 @45 MAINCROP 2. @48 SECONDCR 2. @51 THIRDCR 2.;
```

Here the pointer would be moved to column 10, then to columns 20, 35, 40, 45, etc. Variables may be skipped:

```
INPUT @45 MAINCROP 2.;
```

or reordered:

```
INPUT @45 MAINCROP 2. @10 ID $4 @40 TYPE $1;
```

If, in the long example above, it seems burdensome to skip single spaces by calculating the position of the next variable, rest assured that in SAS there are alternatives. The simplest alternative is the plus (+) pointer instruction, which is the signal to advance the given number of columns. Thus, in the example above, insertion of +1 rather than @48 and @51 would produce the same result and would avoid determining the exact columns. The following are other examples of the use of the plus pointer instruction:

```
DATA PRACTICE;
INPUT IDN $4. +2 AGE 2. +1 SEX $1. +1 SATSCORE 4.;

DATA TESTANX;
INPUT @10 IDN 4. +1 Q1 1. +1 Q2 1. +1 Q3 1. +1 Q4 1. +1 Q5 1.;
```

The plus convention may be used with an implied list to reduce the need for long repetitive input instructions. These two INPUT instructions are identical in information but direct SAS differently:

```
DATA TESTANX;
INPUT @10 IDN 4. +1(Q1-Q5) (+1 1.);

DATA TESTANX;
INPUT @10 IDN 4. +1(Q1-Q5) (1. +1);
```

A maxim of FORTRAN is that a format specification may not exceed the end of the normal 80 columns of a card (as might happen above). Without detailing how, it must be noted that SAS can reach beyond the end of a line (which demonstrates this system's great flexibility). This is an advanced technique and should not be attempted by the beginner. Terminal lines limited to 80 columns or less follow the card rules. However, some terminals do allow longer lines. The simple solution is to limit the length of the line; otherwise, you will have to ask your consultant for help.

The standard way to advance the pointer to the next line is to use the number sign (#). Thus, #4 directs the pointer to the fourth line (or card) of the current observation. The instruction #6 would move the pointer to column 1 of the sixth line. It is not necessary to direct SAS to the next observation. If the crop information in the aforementioned farm survey were on the second record of each observation, we would have this:

```
INPUT IDN $4. STATE $2. COUNTY $2. TOWNSHIP 3. SECTION 3. ACRES 4.1 TYPE $1.
#2 @5 CARDNO $1. +1 MAINCROP 2. +1 SECONDCR 2. THIRDCR 2.;
```

On the second record, the first four columns would be skipped; then the computer would read a variable named CARDNO in column 5, it would skip one column, and then read MAINCROP in columns 7 and 8, and so on.

The number sign can be used to direct pointer movement in both directions, provided illogical instructions are avoided (such as #6 when there is no sixth line). If, in the preceding example, we wished to read ID and STATE, then MAINCROP, followed by ACRES, we would use this INPUT instruction:

```
INPUT IDN $4. STATE $2. #2 @8 MAINCROP 2. @15 ACRES 4.1;
```

There is no limit to the amount of pointer movement allowed. Problems arise, however, when there are multiple lines and only certain ones are read. SAS does not become confused when records are skipped within an observation; but when records at the end are ignored, SAS cannot know they exist. If the farm survey were actually composed of four cards per observation, SAS would read the second observation from the third and fourth cards of the first observation. This problem does not occur if anything is read from the last record or if the last record is simply acknowledged, for example:

```
INPUT IDN $4. STATE $2. #2 @8 MAINCROP 2. @15 ACRES $.1 #4;
```

The #4 instruction informs SAS of the number of lines per observation. There is no need to indicate records early in the sequence—only the total number of records or lines.

By appropriate pointer movements, the same information may be reread as often as needed. The drivers' license numbers in many states are a good example of the utility of the ability of SAS to reread. Such numbers often contain hidden codes as well as the general identification function. The entire number is the identification variable, and its subparts are read as separate variables.

The pointer may also be controlled by the slash (/) convention, much like FORTRAN-based systems. The slash advances the pointer to the first column of the next line. This is a slight convenience in those situations in which the next card is desired but its sequence number is unknown. The each and number signs are better conventions for the beginner; they force a consideration of the precise nature of the information to be input. The slash does not allow SAS to keep track of the number of records, making it awkward to backtrack.

8.2.1   Column Input

Thus far, only formatted input has been discussed. Column input is similar, but it merely names the columns in which a variable is to be found. In this respect, it is like the DATA LIST feature of SPSS. To illustrate, let us look at a study of the voting behavior of U.S. Senators:

```
DATA SEN_VOTE;
INPUT IDN 1-4 PARTY 6 BILLNO 10-15 VOTE 20;
```

or:

```
INPUT IDN 1-4 BILLNO 10-15 PARTY 6 VOTE 20;
```

Variables may be read in any order, and columns may be skipped simply by not mentioning them. If VOTE had been recorded as Y (yea) and N (nay), we would need the dollar sign ($) convention to indicate a character variable.

```
INPUT IDN 1-4 PARTY 6 BILLNO 10-15 VOTE $20;
```

Decimal information is handled by stating the number of decimal places after the column indicator. For example, if the above survey included the percentage of all possible votes that were actually cast, we would have:

```
INPUT IDN 1-4 PARTY 6 BILLNO 10-15 VOTE $20 PERCENTV 22-24 1;
```

Note that a decimal is not included with the 1, indicating one decimal place. The decimal need not be entered in the actual data; when it is entered, the width of the variable should be expanded by one column to correctly accommodate the decimal. However, SAS does honor actual decimals, thereby overriding the INPUT instructions. Remember that this means that illegitimate decimal values may go undetected.

Column input may be combined with pointer instructions for multiple card observations, as in the following:

```
INPUT IDN 1-4 PARTY 6 #2 BILLNO 10 VOTE $20;
```

The same rules for using the number sign (#) convention with formatted input apply to column input. Clearly, the each sign (@) is not needed because the columns are themselves pointer instructions.

One unique advantage of the SAS column input over similar FORTRAN systems is seen in the situation of unfilled columns. If the weight of a subject is of interest and three columns are reserved, as in the following:

```
INPUT WT 10-12;
```

and if the actual value of 98 pounds is entered in columns 10 and 11 and column 12 is blank, then a FORTRAN-based system will translate this as 980 pounds, but SAS will not. In cases where blanks are zeros, SAS has a special format (see the "Pro Tips and Techniques" section).

Column input is a convenient format because SAS actually translates it into formatted input with pointer instructions. For example, in a study involving four variables (ID, height, weight, and GPA), we might have:

```
DATA GPA_WT_C;
INPUT IDN $ 1-4 HT 6-8 WT 10-12 1 GPA 15-17 2;
```

The INPUT statement would be translated by SAS to this:

```
INPUT IDN $ 4. @6 HT 3. @10 WT 3.1 @15 GPA 3.2;
```

SAS is unique in allowing the several styles to be intermixed. Column input, formatted input, and list input are all translated into formatted input, however.

8.2.2   List Input

The list form of input avoids the necessity of determining column locations or writing pointer instructions and formats. It assumes that the data have been prepared with at least one blank column between each variable. For example, we might have

```
DATA PRACTICE;
INPUT IDN Q1 Q2 Q3 Q4 Q5;
```

IDN will be the first variable that SAS finds in the data line, Q1 the next, and so on. This means that if the blank between IDN and Q1 were not present, then the first value would actually be the true ID value, with Q1 appended to it. And the value that is actually Q3 would be considered Q2, and so on down the line of variables. The last value, Q5, would be considered missing in this case. In studies with comparatively few data points, this error would be detected. When the study is larger or when the person performing the analysis is not the person who collected the data or when missing data are to be expected, then this error condition is unlikely to be routinely detected.

SAS is totally dependent on the sequence of blanks to detect each variable. Obviously, this limits the use of this package. When character variables are part of the list input, simply follow each such variable with the dollar sign ($), for example:

```
INPUT IDN NAME $ Q1 Q2 Q3 Q4 Q5;
```

SAS automatically assigns such variables eight characters; larger groups will be truncated, and shorter ones will have zeros appended. This does not interfere with correct processing because the search continues until a blank is encountered.

There are several special features of list input for those situations in which character variables greater than eight characters must be input or in which embedded blanks must be input. These methods are discussed briefly in the "Pro Tips and Techniques" section and at more length in the SAS manual. It is recommended that the underscore be substituted for the embedded blank and that large character variables simply be formatted as mentioned previously.

The preceding advice implies that the various input procedures may be mixed, and indeed they can. When formatted input is mixed with list input, however, a colon is required. It is placed after the variable and is followed by the format itself. (All of the numerous SAS formats are available.) For example, we could have:

```
INPUT IDN NAME :$ 15. Q1 Q2 Q3 Q4 Q5 GPA :3.2;
```

Unlike regular SAS formats, which end with the last specified column, the special formats inserted after the colon end with the next non-blank column. Thus, if the ID above were 4 columns wide followed by a blank in column 5, and if the name were 15 columns wide followed by a blank, the pointer would be set to column 22. If the name were followed by 15 blanks, then SAS would expect Q1 to be in column 26.

There are two additional problems with list input. If a data value or even a single character is dropped, SAS will interpret it as the signal for a new variable. Alternatively, if an extra character is added, it will not be detected. For example, for a study where human infants were weighed (in ounces) every day for seven days, we might have:

```
DATA WTGAIN;
INPUT IDN WT1 WT2 WT3 WT4 WT5 WT6 WT7;
CARDS;
001 96 94 94 97 99 102 106
002 89 88 87 87 89 93 97
003 145 142 143 149 155 159  60
004 48 47 52 55 59 161
```

With list input, SAS would not detect that the last value in data line 003 and the last value in line 004 are improper. The essence of the problem is that SAS has no way of knowing what size to expect for a value. The first error (60 instead of 160) would have been detected if a column or a formatted input procedure had been used, due to the presence of the blank. SAS would have set the value to "missing" and included a message to the user. The second value would have been truncated to 16 with a formatted input but not with list input. Unless the study had included premature newborns, 16 ounces would have been below the minimum value set for the WT variable, and thus it would have been detected as a possible error.

The final difficulty with list input is the treatment of missing values. One and only one symbol denotes a missing value, and that is the period. If the third WT value were missing from data line 002 above, we would have:

```
002 89 88 . 87 89 93 97
```

With list input, this is the only procedure that is accepted; traditional values such as 0, 9, and 99 cannot be used. This method also eliminates the use of special codes for the responses don't know, refused, or not applicable. SAS can handle all of these responses and more, but not with list input.

With all of the problems of list input, we begin to wonder what it is good for. For one thing, list input is very useful for the experienced researcher using a terminal. With terminal data entry, such errors as dropped characters and extra characters are very easy to detect. This detection is very difficult with cards, and the inexperienced researcher is often unsure of what to look for. Nonetheless, there can be no doubt that the list input is the fastest and easiest way to input data.

8.2.3   Miscellaneous Input Features

The end of the user's instructions and data is signaled by a semicolon. When the data must contain semicolons, special instructions are necessary. In this instance, the instruction CARDS4 is substituted for the regular CARDS, and the next record after the data should contain four semicolons in columns 1 through 4. A similar problem occurs when the next statement after the data is so long that the ending semicolon is on a continuation record. SAS will assume that it is still reading data. This assumption may be forestalled by inserting a single record that has a semicolon after the last data record. Indeed, those familiar with SPSS and BMDP should definitely use this method. It will seem natural, since both of those packages use a special end-of-data procedure.

The final point to be made about input concerns the creation of SAS data sets. Each SAS installation may or may not allow the retention of data sets as part of SAS. These sets have many advantages over cards or operating system files. The SAS-79 manual gives the necessary details of such files, but SAS data sets, if available, may be used without knowledge of the details. The alternative to using data sets is to store the data in a personal file (usually a disk) and to send the file to SAS through the proper control instructions whenever a particular analysis is needed.

8.3   Missing Values and SAS

SAS devotes an entire chapter to the topic of missing values. The treatment here will be brief and will serve more as a familiarization than as a detailed account.

The standard missing value in SAS is the period. It need not be left- or right-justified when in column or in formatted input. SAS can find the period at any location within the specified field. (See the previous discussion of list input.) But SAS is not limited to just the period as a missing value; any numeric value may be used by converting it to a missing value. For example, in a five-question survey, we could declare the value 9 as missing:

```
DATA SIMSURVY;
INPUT IDN Q1 Q2 Q3 Q4 Q5;
IF Q1=9 THEN Q1=.;
IF Q2=9 THEN Q2=.;
IF Q3=9 THEN Q3=.;
IF Q4=9 THEN Q4=.;
IF Q5=9 THEN Q5=.;
CARDS;
```

Clearly, nine individual IF statements could have been written to the same end. The advantage of declaring a specific value to be missing is that SAS will keep track of the number of such variables and print a list of where they were found. Often, however, once an observation contains a missing value, it is dropped from the analysis. This is handled differently in each SAS procedure (program), and the details are covered in the manual. Until SAS is totally familiar territory, each procedure should be checked for the way in which it deals with missing values. Any number of numeric values may be declared missing. Each value requires a separate declaration.

Character values may also be declared missing. To do so, a special MISSING statement is inserted into the data process. For example, we could have these statements:

```
DATA SIMSURVY;
MISSING R;
INPUT IDN Q1 Q2 Q3 Q4 Q5;
```

Under these circumstances, both the period and the character R would be considered to be missing values. All the letters and the underscore may be assigned in this way. Several codes may be established as missing for all variables; in the preceding example it is done on a variable-by-variable basis.

This feature of using a special character for a missing value applies only to numeric variables. When the presence or the absence of a character value is to be determined, a period must be added before the character. For example, we could write this:

```
DATA SIMSURVY;
MISSING N;
INPUT IDN 1-4 (Q1-Q5) (+1 1.);
IF Q1=.N THEN DO;
(insert further instructions)
```

With these instructions, SAS will check Q1 for the value .N. On finding this value, it will execute whatever instructions follow the DO. These might be further statements to count the N's or to print special messages. The point is that the special characters require a period before they may be used in other SAS instructions within a MISSING statement.

Thus, we may have the period, numbers, letters and the underscore as missing-value codes. The last routine SAS code to indicate a missing value is the blank itself. Blanks may be used with all input procedures except list input (see Section 8.2.2). The same number of blanks should be inserted as would have appeared in the actual data. If this is not done, there is a high probability that the data will not match the stated structure. In many ways, it is not a positive action to use a blank. Rather, it is an absence, and so many consultants recommend the overt action of the period or of a special code. This is good advice for the less experienced user to follow.

8.4   SAS Procedures

What SPSS calls a task and BMDP calls a program, SAS calls a procedure. There are PROC ANOVA, PROC REG, and so forth. SAS organization makes all procedures potentially available in a single job; i.e., the job is one large program and not a group of smaller programs. Multiple INPUT instructions are also allowed. (These multiple options are discussed in Section 8.7 of this chapter.)

This section is not intended as a review of the abilities of SAS procedure by procedure. Instead, it is intended to introduce the basics of requesting an analysis.

8.4.1   Requesting an SAS Analysis

Requesting an SAS analysis is known as the procedure (PROC) step, just as the sections on input prior to this represent the data (DATA) step. The word PROC should not be confused with the IBM JCL instruction PROC=. They are both abbreviations of the word procedure, but they are not related to each other except in the sense that at some SAS installations one might well write the JCL instruction PROC= in order to use the SAS system. The equals sign is the distinguishing feature. (Recall that JCL stands for IBM's Job Control Language. SAS is currently available only for the IBM equipment.)

The following is an example of a complete SAS analysis:

```
(insert JCL here)
DATA EXAMPLE1;
INPUT IDN 1-4 Q1 6 Q2 7 Q3 8 Q4 10 Q5 11 Q6 12 SEX 14;
AVGQ=(Q1 + Q2 + Q3 + Q4 + Q5 + Q6)/6;
CARDS;
0001 342 566 1
0002 125 799 1
0003 533 767 1
0004 611 758 2
0005 112 569 2
PROC UNIVARIATE FREQ;
ID=IDN;
(insert JCL here)
```

In this case, we can see a number of SAS operations:

1. The data set created is named EXAMPLE1.
2. The data set is on terminal lines (or cards).
3. The data are described to SAS by an INPUT instruction.
4. New variables can be created by manipulating old ones.
5. SAS data records do not require a semicolon.
6. The analysis requested is named UNIVARIATE, and within that analysis the frequency statistic (FREQ) is desired.
7. There is a UNIVARIATE option named ID (it serves a labeling function).

Of course, some SAS analyses are much more sophisticated than this example. Another basic form is seen when the data set is in SAS, as in the following:

```
DATA;
SET EXAMPLE1;
PROC SUMMARY;
CLASS SEX;
VAR ALL;
OUTPUT OUT=EXAM1SUM MEAN RANGE STD;
PROC PRINT N;
```

Here, SAS will create a temporary data set named WORK.DATA1 whose contents will be that of EXAMPLE1. The output of the SUMMARY procedure will be named EXAM1SUM, and it will contain the mean, the range, and the standard deviation (STD) for all variables in EXAMPLE1. The results will be broken down further according to gender (SEX). We also see illustrated the process of passing the output from one procedure step to another for further action. If data set modification had occurred between the PROC statements in this example, the PRINT procedure would print the most recent data set. This can be avoided by writing

```
PROC SUMMARY DATA=SAVE.EXAMPLE1;

PROC PRINT DATA=EXAM1SUM N;
```

The options available to SAS procedures are not wholly consistent; e.g., DATA= (see Section 8.1) is available in the majority of procedures but not in all. SAS procedures can use the following instructions: DROP, KEEP, LABEL, FORMAT, TITLE, COMMENT, MACRO, RUN, and OPTIONS. These, in effect, are utilities that are meant to tailor data sets to each user's needs. (See comments and suggestions in the "Pro Tips and Techniques" section.) The user is strongly advised to review each procedure thoroughly before assuming that a specific feature is available. A quick review of which procedures are available in SAS can be found on the front flyleaf of the SAS manual. This organizational chart also shows some of the relationships between procedures.

8.5   SAS Applied

With the foregoing brief overview and the review of SAS input instructions, you are ready to begin working. Remember that the specifics of each program are in the SAS manual—not in this text. You should be sure to read pages 1–35 and Sections 8, 9, 12, and 13 of Part I of *SAS-79 User's Guide*. Read the manual in conjunction with each applied section that introduces new analysis procedures. The goal is to provide practice using SAS as well as some examples for future reference.

8.5.1   Practice with the FREQ (Frequency) and UNIVARIATE Procedures

The place to begin learning SAS is with the frequency (FREQ) and the UNIVARIATE procedures. Both are descriptive in function, with some degree of overlap. It is not uncommon to need both of these procedures to obtain a complete picture of the data. To learn SAS, you do not have to complete all of the exercises, but it will be instructive to do so.

When reading SAS, remember that many of the facilities are part of the data step. Proper use of SAS depends on adequate manipulation of these data sets. Indeed, at first many of the actual procedures seem to lack options. They are actually present but are hidden in the data step.

=========== **EXERCISE 1** ===========

The highly simplified research question of concern here is whether or not the U.S. voters have shifted politically to the right. One way to answer this is to interview individuals about their past voting behaviors. If large numbers of voters who voted Democratic in the past are now voting Republican, then the thesis that voters have indeed become more conservative is supported. The variables in the study are *possible* variables and are not meant to be definitive of this type of research.

The basic data are given in Table 8.1. Names are deliberately not assigned to variables so that you may practice naming. The goal here is to write statements that will direct SAS to perform a frequency tabulation (FREQ) on all variables of interest. Do not allow SAS to produce frequency tables for all combinations and levels of variables. Practice is the point—not merely the production of paper.

The answer that follows produces a count of sex by vote for each year. Remember that not every voter voted every year, making conventions for missing values especially important here. The number 9 was chosen as the code for ineligibility, 0 indicates eligibility but no vote, and the period indicates totally missing data. (See Appendix A for additional data.) Assume that the data are on terminal lines or cards and that you do not wish to retain the data set in SAS. Further, assume that you wish all statistical calculations to be made (gamma, etc.) and that all variables are to be entered. Use formatted input procedures. Finally, assume that voting is coded R for Republican, D for Democrat, and T for third party.

Answer:

```
DATA VOTESUR;
INPUT IDN 3. @5 IVDATE 6. @12 IVID 2. @16 SEX $1. +1 AGE 2. +1 VOTE76 $1. +1
    VOTE72 $1. +1 VOTE68 $1. +1 VOTE64 $1. +1 VOTE60 $1. +2 REGIS 1. +1 STATE 2. +1
    COUNTY 2. +1 MUN 2. +1 RURALA 2.;
CARDS;
(insert data here)
PROC FREQ;
TABLES SEX*VOTE76 SEX*VOTE72 SEX*VOTE68 SEX*VOTE64 SEX*VOTE60/MISSING ALL;
TITLE PRESIDENTIAL ELECTION VOTING BEHAVIOR;
```

Your answer should be similar to that given; obviously, variable names may differ as well as the INPUT statement, which is for the data as presented in Table 8.1. There are several things that must be noted about these instructions.

Control of the SAS pointer with the each sign (@) (as opposed to the plus sign) is strictly a matter of convenience. In this example, either sign could have been

TABLE 8.1. Data on Presidential Election Voting Behavior

| | | Data Description | |
| | | Number of | |
Variable	Size	decimal places	Function
1	3	0	identification
2	6	0	date of interview
3	2	0	interviewer ID
4	1	0	gender of respondent
5	2	0	age
6	1	0	vote in 1976
7	1	0	vote in 1972
8	1	0	vote in 1968
9	1	0	vote in 1964
10	1	0	vote in 1960
11	1	0	current party registration
12	2	0	state
13	2	0	county
14	2	0	municipality
15	2	0	rural area

Raw Data

```
001   800928 19 M 24 9 9 9 9 9   9 07 20 17 00
002   800927 19 F 29 R R 9 9 9   R 07 09 00 14
003   801002 07 F 41 D D D D D   D 07 09 00 14
004   800930 19 F 37 R 0 D D 9   R 07 09 00 14
005   801004 08 M 36 R R R 9 9   R 07 20 17 00
006   801001 08 F 33 D D 0 9 9   D 07 20 17 00
007   801003 07 0 28 0 9 9 9 9   D 07 20 17 00
008   800929 19 M 31 0 D 9 9 9   D 07 20 17 00
```

used, but the each sign is easier to use when skipping blocks of columns. For example, if the variable SEX had been in column 72, we could have written this:

```
INPUT IDN 3. @5 IVDATE 6. @12 IVID 2. @72 SEX $1. +1 AGE 2. +1 . . .
```

It is not necessary to read SEX as the fourth variable. Recall that control of the pointer with the each sign allows any order. The plus sign does not allow for backtracking, e.g., a -6 could not be used. It is possible to reverse the pointer with the plus convention, but it is not a recommended procedure. (See page 63 of the SAS manual.)

The INPUT instruction given in the answer can be simplified by using the implied-list feature (Q1-Q5, and so on). SAS, though, is not restricted to sequential variables but only to contiguous variables that have been previously defined. For example, the following *does not* function correctly:

```
INPUT IDN 3. @5 IVDATE 6. @12 IVID 2. @16 SEX $1. +1 AGE 2.
  VOTE76-VOTE60(+1 2.) +1 REGIS 1. +1 STATE-RURALA(+1 2.);
```

The first implied list has the correct form, but it implies 16 variables—not 5. The STATE-RURALA list is incorrect for two reasons. First, the proper form for variables not ending with a sequence number is two consecutive minus signs (STATE--RURALA). In addition, there is no way for SAS to know that COUNTY and MUN exist. Finally, implied lists must be placed within parentheses when used in input. Here is the correct form for the above INPUT instruction:

```
INPUT IDN 3. @5 IVDATE 6. @12 IVID 2. @16 SEX $1. +1 AGE 2.
  (VOTE1-VOTE5)(+1 $1.) +1 REGIS 1. +1 STATE 2. +1 COUNTY 2. +1 MUN 2. +1
  RURALA 2.;
```

Assigning variable names is important to proper analysis. Therefore, the list feature should be used in SAS INPUT statements only with sequentially numbered variables. Once variables are named, the use of the list feature does not create similar problems. To illustrate this, consider the TABLES statement of this answer. It could have been written as follows:

```
TABLES SEX*VOTE76--VOTE60/MISSING ALL;
```

In this instance, SAS already knows the names of the variables between VOTE76 and VOTE60. All resulting tables will be properly labeled. This TABLES statement is an example of a situation that is easily confused when using SAS. Implied lists in INPUT statements require one minus (−) to separate the first and the last elements. All other implied lists use two minus signs (−−). Both cases require parentheses.

The remainder of the answer offers no special surprises. It requests the frequency procedure (FREQ) for five, two-way tables, assigns the period as the missing variable, and requests additional statistics. One-way tables could be requested at the same time, as in the following:

```
TABLES STATE SEX*VOTE76--VOTE60/MISSING ALL;
```

SAS does not impose a restriction on the number or combinations of tables. This lack of limitation can produce serious overkill when it results in volumes of tables. The tables may also be used as a check on the accuracy of data production. Consider the following instruction:

```
TABLES SEX;
```

If the resulting tables gave the frequency of anything other than men (M), women (W), and missing (.), then bad data would be found somewhere. Note that when

SAS encounters a number when a character was expected (or vice versa), it will tell you so. However, if you had punched an N instead of an M in the data, SAS would be unable to detect this without special programming (see Exercises 3 and 4). The FREQ procedure is a quick and clever way to check this, and it is highly recommended.

The frequency procedure often does not provide a complete selection of descriptive statistics. The beginner is advised either to practice with Exercise 2 or to use Exercise 1 with the UNIVARIATE procedure. Discussion of this procedure can be found in Exercise 2.

═══════ EXERCISE 2 ══

Whether or not thermal biofeedback can prevent migraine headache is the research question for this exercise. There are two important independent variables: the type of headache and the experimental condition. Migraine headaches (for this exercise) are divided into the type in which the onset of pain is preceded by a drop in the temperature of the hands and the type in which there is no temperature drop. The four experimental conditions are thermal biofeedback training, false thermal training, forearm-muscle relaxation training, and control (no training). The treatment group is to be compared to the other three control variations for each type of headache. The dependent measure is the number of headaches before treatment, immediately after treatment, and after six months (average duration of headaches is of secondary interest). Remember that the purpose here is to practice with SAS, and not to become overly involved in the details of experimental design.

The basic data are given in Table 8.2. Variables are noted by function rather than by name so that you may practice naming. The raw data need not be arranged as implied, but the answer given is based on this arrangement. The specific assignment is to write the statements needed to direct SAS to perform a UNIVARIATE analysis. Use frequency statements to check for improperly prepared data. Conceptually, these analyses would serve as a preliminary check of the data.

Use the period as the missing values convention. UNIVARIATE does not analyze nonnumeric variables, and thus the gender codes must be modified or the FREQ procedure must be used in order to check this variable. Both methods will be discussed. Lastly, obtain the plots of age, the pre-stress index, and the number of migraines before treatment. Assume that the data are on terminal lines or cards and that the data set is not to be retained. (See Appendix A for additional data.)

Answer:

```
DATA HEADACHE;
INPUT IDN 4. @10 SEX $1. +1 AGE 2. +1 HEADTYPE 1. +1 EXCONDIT 1. +2 PREACHE 2.
 +1 PREDUR 4. +1 PSTACHE 2. +1 PSTDUR 4. +1 SIXACHE 2. +1 SIXDUR 3. +2
 PRESTRES 1. +1 PSTSTRES 1.;
```

```
CARDS;
(insert data here)
PROC UNIVARIATE;
VARIABLE IDN AGE HEADTYPE EXCONDIT PREACHE PREDUR PSTACHE PSTDUR SIXACHE SIXDUR
 PRESTRES PSTSTRES;
PROC UNIVARIATE PLOT;
VARIABLE AGE PRESTRES PREACHE;
PROC FREQ;
VAR SEX;
```

This answer simply and clearly names the data set HEADACHE and then informs SAS what the variables are and where they are located. The only character variable is SEX. Note that SAS statements may begin in any column and that in

TABLE 8.2. *Data on Migraine-Headache Reduction through Thermal Biofeedback*

		Data Description	
Variable	*Size*	*Number of decimal places*	*Function*
1	4	0	identification
2	1	0	gender
3	2	0	age
4	1	0	headache type
5	1	0	experimental condition
6	2	0	headaches/month before treatment
7	4	2	average duration in hours:min.
8	2	0	headaches/month after treatment
9	4	2	average duration in hours:min.
10	2	0	follow-up: headaches/month at six months after treatment
11	4	0	average duration in hours:min.
12	1	0	pre-stress index
13	1	0	post-stress index

Raw Data

```
0001    F 39 1 1    04 0820 01 0210 01 150 3 4
0002    F 22 2 1    07 0530 00 0000 01 050 5 2
0003    F 40 1 2    06 0400 06 0510 05 400 7 8
0004    F 28 2 2    03 0700 03 0600 03 300 5 4
0005    F 37 1 3    07 0210 05 0500 04 430 6 3
0006    F 36 2 3    04 0630 03 0400 04 400 7 7
0007    F 24 1 4    05 0530 06 0500 04 606 6 3
0008    F 29 2 4    04 0430 05 0530 05 530 4 6
```

multiline situations the break between lines should be between variables. The decision to use +1 instead of the each sign (@) is one of convenience; it could have been done the other way.

The first PROC statement eliminates SEX as a variable by naming all of the variables except SEX. Since UNIVARIATE works only with numeric variables, SEX would cause an error message. However, the instructions for this exercise request a check of all variables. The answer resolves this by requesting another procedure (in this case FREQ), and by limiting it to the SEX variable. Recall from Exercise 1 that FREQ provides descriptive information about both numeric and character variables. This does not imply that FREQ and UNIVARIATE provide the same information, but it does imply that both can provide preliminary data checking.

When variables are unsuitable for analysis by a specific procedure, there is not always an alternative procedure. The solution then is to create new variables from the old ones. SEX, in the present case, would be suitable for UNIVARIATE if it were numeric. To perform this substitution with SAS, we are required to use data manipulation statements, for example:

```
DATA HEADACHE;
INPUT (same input as answer)
NSEX=0.;
IF SEX='F' THEN NSEX=1;
IF SEX='M' THEN NSEX=2;
CARDS;
PROC UNIVARIATE;
```

The data set created by SAS will now include a new variable, NSEX, coded 1 for females and 2 for males. The output from UNIVARIATE will include 14 variables instead of the original 13. As stated above, error messages will still be produced if SEX is included. This can be avoided by using the variable (VAR) option in the PROC step and by specifying all variables except SEX. Clearly, NSEX would be included.

Note that nothing specific is stated in this answer about missing values, so the assumed value is in effect (recall that the period is the assumed missing value). UNIVARIATE automatically counts the missing values and prints the results as NMISS. A missing value affects each variable separately; in FREQ, a missing value eliminates the complete observation from the tabulations. It is a good idea to note the particular way each SAS procedure treats missing values before starting an analysis.

The final assignment for this exercise was to obtain plots of selected variables. Obviously, specifying PLOT in the first procedure request would be sufficient, but that would produce a plot for every variable. The number of unwanted plots can be reduced by requesting another UNIVARIATE having only the three desired variables. Conversely, this additional UNIVARIATE request also will repeat the calculations for the three variables. In general, the additional PROC request is preferred if for no other reason than it keeps the amount of paper produced under control.

A final comment on the use of the multiple procedure request is in order. SAS works with the most recently established data set unless directed otherwise in a DATA= statement. When the data are modified between procedure steps, subsequent analyses are performed on the most recently created set. It takes a data step to create a data set, and thus if we request a particular set in a PROC, it is not the most recent set. Many SAS procedures allow the creation of an output data set as an option. When this is done, it becomes the most recently created data set. Exercise caution when many successive procedure steps are used.

8.5.2   Practice with the CORR (Correlation) and TTEST Procedures

The two exercises that follow do not overlap as much as do Exercises 1 and 2. The experienced researcher should pick and choose between them. The beginner and those who are experienced with other packages but new to SAS should work through both exercises because they approach the data differently. Remember that these exercises are not a substitute for reading the SAS manual.

===== **EXERCISE 3** =====

Utilizing the data in Exercise 1, we now consider the following study. One way to determine if voting behaviors have shifted over the years is to obtain a correlation between the earliest vote and the most recent. Since the data are restricted as to value (mostly R's and D's), a Pearson correlation or a Spearman rank correlation would be inappropriate. Therefore, obtain a correlation between VOTE76 and VOTE60 using Kendall's (tau-*b*) concordance procedure. An interesting extension is to obtain a matrix of concordances for all voting years. The logic is that if there were no shifts whatsoever between the years, then the concordance rate would be 1.00; and if every voter changed his or her behavior, it would be 0.00. Use the correlation (CORR) procedure for both of these tasks.

The last task is to obtain a Pearson correlation between AGE and a new variable, AVGVOTE, arranged by SEX. To accomplish this, the data set must first be sorted according to the variable SEX (use the procedure SORT) and then must be analyzed by the procedure CORR. The variable AVGVOTE must also be determined. Do this by writing statements that sum the five votes and then divide them by five. While R's and D's cannot be divided by any number, each party can be considered a point on the conservatism/liberalism continuum, and the numbers 1 and 2 can be assigned to each of those persuasions, respectively. It could be argued that this is a violation of the Pearson correlation assumption, but for the purposes of this exercise we can assume that the numeric properties of the new variable (AVGVOTE) are adequate.

Assume that the data are on terminal lines or cards and that a special output matrix is not desired. Carefully read the SAS instructions on choosing the three available correlations. Also assume that all calculations should be performed on observations without missing data.

Answer:

```
DATA VOTECORR;
INPUT IDN 3. @16 SEX $1. +1 AGE 2. +1 VOTE76 $1. +1 VOTE 72 $1. +1
  VOTE68 $1. +1 VOTE64 $1. +1 VOTE60 $1.;
IF VOTE76='D' THEN NVOTE76=1;
IF VOTE76='R' THEN NVOTE76=2;
IF VOTE72='D' THEN NVOTE72=1;
IF VOTE72='R' THEN NVOTE72=2;
IF VOTE68='D' THEN NVOTE68=1;
IF VOTE68='R' THEN NVOTE68=2;
IF VOTE64='D' THEN NVOTE64=1;
IF VOTE64='R' THEN NVOTE64=2;
IF VOTE60='D' THEN NVOTE60=1;
IF VOTE60='R' THEN NVOTE60=2;
AVGVOTE=(NVOTE76+NVOTE72+NVOTE68+NVOTE64+NVOTE60)/5;
CARDS;
(insert data here)
PROC CORR KENDALL NOMISS;
VAR VOTE76 VOTE60;
TITLE VOTE IN 76 CORRELATED WITH VOTE IN 60;
PROC CORR KENDALL NOMISS;
VAR VOTE76--VOTE60;
TITLE INTERCORRELATIONS OF ALL VOTES;
PROC SORT;
BY SEX;
PROC CORR NOMISS;
VAR AVGVOTE AGE;
BY SEX;
TITLE CORRELATION BETWEEN 5 YEAR AVGVOTE AND AGE BY SEX;
```

The first observation to be made about this answer concerns the multiple procedure steps. Except for the PROC SORT, they could have been merged into one request—at the cost of the descriptive titles, however. Since both Pearson and Kendall analyses are needed, a single PROC would of necessity have had to call for both, thereby generating a number of unwanted and wasteful analyses (for example, a Pearson analysis for VOTE76 and VOTE60). Note that making separate requests tends to reduce confusion. When an error exists in a single large PROC, SAS will not complete any of the analyses; however, when an error is in one of several PROC instructions, only that one is not completed.

The SORT request could have come at any point in the program control language (PCL) as long as it preceded any PROC that included the BY SEX request. In this instance, it will not affect the other requests to sort by SEX.

This answer demonstrates passing the same data from one procedure to the next. Changes made at any point continue to the next request. As in this example, once the data set is modified, it is modified for all subsequent analyses. When changes are necessary for some analyses and not for others, separate data sets must be established. If for some reason the new variable (AVGVOTE) should have been present for only the first procedure, there are two solutions.

The first approach is to reverse the order of the procedures, and then to create a new data set for the needed analysis prior to the last PROC. For example, we could have constructed this answer as follows:

```
DATA VOTECORR;
INPUT (same input as before)
CARDS;
(insert data here)
PROC CORR KENDALL NOMISS;
VAR VOTE76 VOTE60;
PROC CORR KENDALL NOMISS;
VAR VOTE76-VOTE60;
DATA VOTECOR2;
SET VOTECORR;
IF VOTE76='D' THEN NVOTE=1;
(continue IF instructions as before)
AVGVOTE= (as before)
PROC SORT;
BY SEX;
PROC CORR KENDALL NOMISS;
VAR AVGVOTE AGE;
BY SEX;
```

The new data set (VOTECOR2) will have the same information as the old (VOTE-CORR), in fact it will be an exact copy. Then the variable AVGVOTE is calculated and added to the new data set. It is not necessary to invent a new name for the new data set. If none is given, SAS will assign a temporary name, _DATA_ in this case. Whenever a change is temporary and is not to be retained, this name is sufficient. The temporary name _NULL_ could also be used in this case. Obviously, when many sets are being processed concurrently, it is wise to establish mnemonic names as an aid to maintaining order.

The second approach to data sets with unwanted variables is to use the DROP instruction in the data step. Simply establish a new set from the old (as above), and then drop the unwanted variables. Some SAS procedures also allow the DROP statement. When this is possible, the DROP statement functions in the same way as the DROP instruction in the data step, but it avoids the necessity of a new data step.

Another lesson from this answer is seen in the statements that create the new variable, AVGVOTE. SAS has a complete line of manipulations of this sort: all the usual arithmetic operations, the trigonometric functions, logical operators (e.g., not equal to), and more. Working examples of some of these appear in the remaining exercises, and they are mentioned in the "Pro Tips and Techniques" section.

In the data manipulation of this answer, we can see that SAS allows somewhat complex statements of the IF . . . THEN . . . variety. Their complexity is due to the fact that the dependent parts of such expressions can themselves be expressions. Consider the following:

```
IF X > Y THEN X=X/2;
```

This expression will adjust all values of X that are greater than Y by dividing X by 2 and using the result as the new value of X. Each observation is so modified.

In this answer, the five variables VOTE76 through VOTE60 are tested to see if they equal the character D. If so, the new variables (NVOTE76, etc.) are set equal to 1. Similar logic applies to determining the R's. Although it is not efficient to write individual statements, this method avoids ambiguities.

It is better for beginning and intermediate users to write conversions that are inefficient but that work the first time. You will find more efficient methods in both of the remaining exercises and in the discussion of syntax.

═══════ EXERCISE 4 ═══════

The data and the design from Exercise 2 are analyzed in a different way in this exercise. We see an interesting experimental comparison between those subjects receiving the experimental treatment (thermal biofeedback) and the placebo group (forearm-muscle relaxation training). To make this same comparison using a t-test, the SAS TTEST procedure is the obvious choice. TTEST requires that a CLASS= variable be declared in order to form the groups. This variable *cannot* have more than two values. Since there are four groups, you must devise a means of reducing the number of values. There are a variety of ways to accomplish this. The best method is to use the special data set manipulation ability of SAS. (The answer given is based on this ability.)

As a second problem, obtain a t-test comparing the experimental groups and each of the three control groups. Again, the creation of special data sets with the desired properties is the preferred solution.

In both of these problems, the post-treatment number of headaches should be made the dependent measure. When the analysis task is stated in this way, it implies that this is a legitimate analysis. In reality, multiple t-tests are not the best testing method. It is also not a good idea to test only two groups within a complex of groups. These questionable analyses are done here solely to illustrate the SAS technique in a manageable way. Assume that the data are on terminal lines or cards.

Answer:

```
DATA HEADACHE;
INPUT IDN 4. @10 SEX $1. +1 AGE 2. +1 HEADTYPE 1. +1 EXCONDIT 1. +2 PREACHE
   2. +1 PREDUR 4. +1 PSTACHE 2. +1 PSTDUR 4. +1 SIXACHE 2. +1 SIXDUR 3. +2
   PRESTRES 1. +1 PSTSTRES 1.;
CARDS;
(insert data here)
DATA TEST1;
```

```
SET HEADACHE;
IF EXCONDIT LE 2;
PROC TTEST;
CLASS EXCONDIT;
VAR PSTACHE;
TITLE T-TEST BETWEEN GROUPS 1 AND 2;
DATA TEST2;
SET HEADACHE;
IF EXCONDIT=2 THEN DELETE;
IF EXCONDIT=4 THEN DELETE;
PROC TTEST;
VAR PSTACHE;
CLASS EXCONDIT;
TITLE T-TEST BETWEEN GROUPS 1 AND 3;
DATA TEST3;
SET HEADACHE;
IF EXCONDIT=2 THEN DELETE;
IF EXCONDIT=3 THEN DELETE;
PROC TTEST;
VAR PSTACHE;
CLASS EXCONDIT;
TITLE T-TEST BETWEEN GROUPS 1 AND 4;
```

In this answer, note that the INPUT instruction is the same as that in Exercise 3. You may wonder whether this is superfluous and wasteful of time and storage. Simply put, it is. For the purposes of this problem, the following line would have sufficed:

```
INPUT IDN 4. @17 EXCONDIT @28 PSTACHE 2.;
```

In those circumstances where writing a new INPUT instruction is more trouble than it is worth, space and time may be conserved by using the DROP instruction. We could write, for example:

```
DATA HEADACHE;
INPUT IDN 4. @10 SEX $1. +1 AGE 2. +1 HEADTYPE 1. +1 EXCONDIT 1. +2
  PREACHE 2. +1 PREDUR 4. +1 PSTACHE 2. +1 PSTDUR 4. +1 SIXACHE 2. +1
  SIXDUR 3. +2 PRESTRES 1. +1 PSTSTRES 1.;
DROP SEX AGE HEADTYPE PREACHE PREDUR PSTDUR SIXACHE SIXDUR PRESTRES PSTSTRES;
CARDS;
```

Here, the data set HEADACHE will contain only the variables IDN, EXCONDIT, and PSTACHE, which reduces its size considerably and therefore speeds up all of the analyses. The DROP instruction may be used in both the data and the procedure steps. Only small storage savings will be realized if DROP is used in the PROC step. *Note:* the KEEP instruction (the opposite of DROP) would work more efficiently in this example.

Another aspect of SAS that is observable in this answer is the ease with which data sets are created and manipulated. If the steps TEST1, TEST2, etc., had not been named, then SAS would have assigned the names DATA1, DATA2, etc. It is usually better to assign names because it makes tracing errors easier. The name _NULL_ could also have been assigned to each of these later data steps, if the information did not need to be retained.

In the second data step, we can observe the procedure for substituting a data set. The TEST1 set will not contain all values of EXCONDIT as a result of the IF instruction. The new set TEST1 will be comprised of a portion of the set HEAD-ACHE. The portion will be determined by the logic of the IF statement. In this case, an observation is placed in the set if (and only if) the value of variable EX-CONDIT is less than or equal to 2. The full range of arithmetic and logic statements is available (see Appendix 5 in the SAS-79 manual). The function of the IF state-ments is to limit the variable EXCONDIT to three values: 0, 1, and 2. Since only 1's and 2's are present, there is no 0 group. There are, of course, several ways to obtain this logic.

The purpose of breaking the IF into subsets is to satisfy the requirement of the TTEST procedure. The TTEST must have a CLASS= instruction having no more than two values. In this example, there are four experimental conditions requiring these manipulations.

Much the same logic can be seen in the second and third t-tests. Again, a variety of logical approaches could be used to select the values.

Finally, note that the subparts of the procedure steps are stated on separate lines. Separate lines are not required but are convenient for those who are experi-enced only with SPSS and BMDP. Separate lines make for easier removal or substi-tution when working with cards. If you are working with terminal lines, this is not as important.

8.5.3   Practice with the CHART and PLOT Procedures

The CHART and PLOT procedures are valuable tools for visualizing the relationships within and between variables. In a broad way, CHART applies mostly to variables that are discrete (or at least they are thought of as being discrete). PLOT, on the other hand, requires continuous variables. An example of a variable that can be conceived of as either discrete or continuous is age in a longitudinal study of human develop-ment. A bar chart showing age versus size of vocabulary could be very enlightening if it covered a small number of ages (or age ranges). How-ever, if this analysis were done on a monthly basis up to age 21, there would be 252 individual bars with which to contend. When age is con-ceived of as a continuous variable, it allows us to use scaling and some other mathematical operations to reduce the problem to a manageable size.

A word of caution is in order here. It is not legitimate to pretend that a discrete variable is continuous. The PLOT procedure does not change the nature of the variable. Discrete methods may be used with continuous data, but the reverse is not true.

Use the data and variables from Exercise 1 to obtain frequency charts for each of the five presidential voting years. The information that appears on the horizontal axis should be the two types of votes—Democratic and Republican. (Missing values and other codes would simply add more categories to the horizontal axis.)

Next, examine the years 1976 and 1960 as above, but add the SEX variable. You should then have a bar graph for each year showing the joint frequency of party and gender. If you are feeling adventurous, you can also produce cumulative frequencies and even pie charts. Note that the frequency option does not apply to this exercise.

Finally, sum the variable AGE and chart it against VOTE76 for Democrats and for Republicans.

Answer:

```
DATA VOTECORR;
INPUT IDN 3. @16 SEX $1. +1 AGE 2. +1 VOTE76 $1. +1 VOTE72 $1. +1
  VOTE68 $1. +1 VOTE64 $1. +1 VOTE60 $1.;
CARDS;
(insert data here)
PROC CHART;
VBAR VOTE76--VOTE60;
VBAR VOTE76 VOTE60/SUBGROUP=SEX;
PROC CHART;
VBAR VOTE76/SUMVAR=AGE;
```

The INPUT instruction of this answer is a shortened version of that given in Exercise 1, but, as we mentioned before, it is wasteful to create data sets that are larger than necessary. We can also see a typical multiple PROC setup, although in this case the multiple VBAR instructions are acceptable.

The first CHART request results in five separate bar graphs, each having a voting year on the horizontal axis and the frequency of R's and D's on the vertical axis. The labeling of the vertical axis is determined by SAS, as is the scaling of continuous variables. SAS uses the values it finds for each variable, but a simple R or D is not very informative. Expanded labels are created by yet another procedure—FORMAT. For example, we might have

```
PROC FORMAT;
VALUE PARTFMT D=DEMOCRAT R=REPUBLICAN;
PROC CHARTS;
VBAR VOTE76--VOTE60;
FORMAT VOTE76--VOTE60 PARTFMT;
```

This would substitute the full labels in the output from CHART. Although this process appears simple, there is some complexity involved. Be very familiar with SAS before adding this technique to your repertoire of skills.

The second VBAR produces two bar graphs that have the variable SEX coded into each bar; i.e., each bar will be composed of F's and M's. Some caution should be exercised in designating the subgroups of the bars. If some chosen variable has 20 different values, the resulting bar graphs will be difficult to read. Three or four variables is about the maximum.

The final CHART request calls for VOTE76 on the horizontal axis and the summation of AGE on the vertical axis. Thus, it allows a side-by-side comparison of the number of Republican to the number of Democratic voters in 1976. If STATE had been included in the data set, then this comparison could be made state by state, as in the following:

```
PROC CHART;
VBAR VOTE76/SUMVAR=AGE GROUP=STATE;
```

If the data are extensive (e.g., all 50 states), then 50 charts, on 50 pages, will be the result of this simple instruction. With census data, such a statement would be disastrous.

The SAS-79 manual details many other types of charts that can be created.

═══════ EXERCISE 6 ═══════════════════════════════════

Using the data from Exercise 2, obtain a plot of pre-headache duration versus post-headache duration for each treatment group. Then form a single plot combining the pre-duration and the post-duration data from all four groups. Allow SAS to define the axes. Also, modify both axes to start near the actual lowest durations. Lastly, calculate two new variables: the difference between the pre- and the post-durations and the difference between the number of pre- and post-headaches for each treatment group. Plot these new variables against each other. Use only whole hours in these calculations.

Answer:

```
DATA HEADACHE;
INPUT IDN 4. @10 SEX $1. +1 AGE 2. +3 EXCONDIT 1. +2
  PREACHE 2. +1 PREDUR 2. +3 PSTACHE 2. +1 PSTDUR 2.;
DELTADUR=PREDUR-PSTDUR;
DELTAACH=PREACHE-PSTACHE;
PROC SORT;
BY EXCONDIT;
PROC PLOT UNIFORM;
BY EXCONDIT;
PLOT PSTDUR*PREDUR;
PROC PLOT;
```

```
BY EXCONDIT;
PLOT DELTADUR*DELTAACH;
PROC PLOT;
PLOT PSTDUR*PREDUR=EXCONDIT;
```

The INPUT statement here is the same as those in Exercises 2 and 4 except that the format for PREDUR and PSTDUR is reduced by two columns to eliminate the minutes. This is simpler than using multiple arithmetic statements. The SAS UPDATE function could also be used, but it requires some familiarity with file-processing operations.

The two difference scores DELTADUR and DELTAACH are calculated for each observation. SAS will note that the number of variables has increased by two. The entire range of arithmetic, algebraic, and logic operations is available. In the syntax of such statements, the logic must not only be correct; it must result in operations that SAS can perform. Section 8.6 attempts to point out certain problems that are common with SAS. As a rule, the logic is easier to follow and is therefore more likely to be correct if we use many short statements rather than a few very complicated ones.

The first procedure step is a request to sort the data by EXCONDIT. This does not change the data in any way, but it does make possible the reorganization of the succeeding PLOT requests in terms of the variable EXCONDIT. Thus, the researcher can compare the scores for each group. This is a quick way to check for outliers or for otherwise strange data. When a later PROC does not request a BY organization, the result is as if the SORT had not been performed.

The last two procedure steps were not combined into one because to do so would result in a very peculiar syntax. Requesting EXCONDIT in the last PROC causes SAS to print the group variable codes on the plot itself. This is good for visually observing and for clustering. The third PROC has a BY organization, so it will produce a plot for each value of EXCONDIT. If SAS could execute this combination (and it probably cannot), the logic would be as follows: organize four plots for the four values of EXCONDIT, and then plot PSTDUR versus PREDUR using all values of EXCONDIT. But EXCONDIT would have only one value for each BY, hence, we would have four plots—one for each condition—instead of one plot with all of the conditions on it. This may sound confusing, but it illustrates that procedure steps may be organized in a variety of equivalent ways. However, within a PROC, a BY option alters that particular procedure. The best advice is, again, to use many individual procedure steps rather than to combine them and be faced with logic so convoluted that it cannot be thought through.

Note that in Exercise 6 nothing was said about the size and the form of the plot axes. SAS allows the user to specify axis dimensions. Make sure that the range of the specification is generous enough to include all values.

8.5.4   Practice with the STEPWISE and ANOVA Procedures

Exercise 7 demonstrates the STEPWISE procedure for obtaining a regression. SAS allows several other types of regression. However, they are either somewhat limited in scope or very dependent on advanced statistical skills (RSQUARE and GLM are two examples). STEPWISE regression is very straightforward and clear.

The ANOVA procedure is most notable for what it is not. Nested, repeated-measures, and unequal *n*'s designs are all handled by other SAS procedures. Exceptions and qualifications abound in this and other SAS experimental analysis procedures. Careful attention is necessary to ensure that the correct design is implemented. Remember that SAS cannot determine whether or not assumptions and design parameters are met. Beyond these qualifications, ANOVA is a breeze to use.

======= **EXERCISE 7** =======

Using the data again from Exercise 1, obtain a forward selection regression with the vote in 1976 as the dependent measure (the measure to be predicted). Independent variables may be any that you suspect are predictive. (The answer given uses the variables of gender, age, votes in 1972 through 1960, and state and county.) Use the default values for entry and for retention of variables. Calculate a new variable that is the average vote from 1972 through 1960, and add it to the list of dependent variables. Use the DO loop technique for this calculation.

If you are using your own data, pay particular attention to the way missing data are handled. If it is desirable to actually run the answer, the eight respondents given in Exercise 1's voting-behavior survey will not produce very meaningful results. The expanded data set for these exercises (Appendix A) will provide a more meaningful analysis.

As a final task, write an INPUT statement that uses both list and formatted input. The original input from Exercise 1 would be sufficient, of course.

Answer:

```
DATA VOTESUR;
INPUT IDN 3. @16 SEX $ 1. +1 AGE 2. +1 VOTE76 $ VOTE72 $ VOTE68 $
  VOTE64 $ VOTE60 $ +3 STATE COUNTY;
DATA NEWVOTE1;
SET VOTESUR;
ARRAY VOTE(I) VOTE76 VOTE72 VOTE68 VOTE64 VOTE60 NVOTE;
ARRAY NVOTE(I) NVOTE76 NVOTE72 NVOTE68 NVOTE64 NVOTE60;
DO I=1 TO 5;
IF VOTE='D' THEN NVOTE=1;
IF VOTE='R' THEN NVOTE=2;
END;
```

```
AVGVOTE=(NVOTE72+NVOTE68+NVOTE64+NVOTE60)/4;
CARDS;
(insert data here)
PROC STEPWISE;
MODEL NVOTE76=SEX AGE NVOTE72 NVOTE68 NVOTE64 NVOTE60 AVGVOTE
  STATE COUNTY/FORWARD;
TITLE STEPWISE REGRESSION ON 1976 VOTING BEHAVIOR;
```

The first lesson from this answer is a demonstration in mixing various formats. The list in the INPUT instruction does not show much of a savings over previous ones because each variable must have a dollar sign ($) appended to indicate its character status. List input is most efficient with very regular numeric variables (STATE and COUNTY, in this example). For a discussion of list input involving more than 80 characters, see the "Pro Tips and Techniques" section. The remainder of the input is as expected.

We see that immediately after the data input a new data set is defined via a DATA statement followed by a SET instruction. These instructions effectively create a new data set named NEWVOTE1, which comprises the information from VOTESUR. Thus, when the data are modified, the old data remain intact in data set VOTESUR and the new data are in the NEWVOTE1 set only. Thus, a later PROC would analyze the old variables without starting over.

The statements that convert VOTE76 through VOTE60 to numeric variables follow. Clearly, such loops can be a significant savings over a long series of individual IF tests. There is no practical limit to these DO loops, but compound indexes can become confusing. Beginners should make their changes in reasonable steps. There are several subtle points about this conversion of R's and D's to numeric values. First, when a series of variables is indexed, the ARRAY statement is used. Second, SAS uses implicit indexes; e.g., VOTE and NVOTE are incremented by the index I despite the fact that the I is unstated. Third, when a variable is defined as numeric or as character in the INPUT statement, it is permanent. Thus, the new variable NVOTE is given the value of VOTE in terms of 1's and 2's. All of these points are dealt with in more depth in Section 8.6.

The AVGVOTE variable could have been calculated as follows:

```
AVGVOTE=SUM(OF VOTE72 VOTE68 VOTE64 VOTE60)/4;
```

or:

```
AVGVOTE=SUM(OF VOTE72--VOTE60)/4;
```

There is no particular advantage to this, but it does point out that SAS can calculate certain statistics without a special PROC. The statistics called for by PROC MEANS are for observations and are not to be applied across variables as in this case. Remember that SUM and other statistics are performed on the variables contained within the parentheses for the current observation. Stating MEAN (AGE) will not produce the average age; rather, it produces the average of the current observation, which will of course be itself. Its real value is in producing scale scores, etc. (see the manual for SAS-79, p. 444).

The STEPWISE regression procedure is the height of simplicity. One needs only to name the procedure, to write the model, and to select the desired options. If one asks for all of the options in hopes of selecting the best regression, confusion will probably follow. Unless your level of skill allows, pick your style of regression and stick with it. Remember that a STEPWISE regression is not necessarily the end of the line. Many experienced workers use it to gain some insight into the data and then move to more definitive procedures—GLM, for instance.

====== EXERCISE 8 ======

Using the setup from Exercise 2, analyze the data with the ANOVA procedure. Treat it either as a one-way design having four treatment groups (thermal feedback, false thermal feedback, forearm-muscle relaxation training, and no training) or as a two-way design having pre-treatment stress levels as the second measure. In both designs, the dependent measure is the number of migraine headaches that occur post-treatment. Assume that stress has five levels.

A word of caution about the assumptions in this analysis: in many situations, the number of headaches would be obtained through self-reports by the subjects. Self-report data are notorious for not meeting the assumptions of interval measurement. For the purpose of this example, we will assume that the number of headaches has been confirmed independently. Stress assessment often suffers from the same problems, but here we will assume it to be adequately assessed. Include interactions as components of your model.

As a second problem, obtain a one-way analysis of pre–post change scores, i.e., pre-treatment to post-treatment differences in the four experimental groups. This requires the calculation of a new variable and its use as the new dependent variable. Remove negative values from the difference score by setting them equal to zero. Remember that repeated analysis of experimental data is often considered improper and is done here only for illustration.

Answer:

```
DATA HEADACHE;
INPUT IDN 4. @17 EXCONDIT 1. +2 PREACHE 2. +06 PSTACHE 2. +14 PRESTRES 1.;
DIFACHE=PREACHE-PSTACHE;
IF DIFACHE LT 0 THEN DIFACHE=0;
CARDS;
(insert data here)
PROC ANOVA;
CLASS EXCONDIT;
MODEL PSTACHE=EXCONDIT;
PROC ANOVA;
CLASSES EXCONDIT PRESTRES;
```

```
MODEL PSTACHE=EXCONDIT PRESTRES EXCONDIT*PRESTRES;
PROC ANOVA;
CLASS EXCONDIT;
MODEL DIFACHE=EXCONDIT;
```

The INPUT statement here is a shortened version of that used throughout the even-numbered exercises. Either the plus or each sign convention may be used. Since the data are scattered on the cards (or lines), list input is not possible (see Exercise 7 for list input practice).

The second part of the data step creates the difference variable DIFACHE by subtracting pre- and post-values. This variable is then tested for negative values, and, if any are found, they are set equal to zero. The logical LE (less than or equal to) also could have been used, since zero values would merely be reset to zero, which would be proper. The absolute value function would not have resulted in the same actions; the absolute value of −5 is 5, whereas the logic of the answer would replace −5 with 0. This is noteworthy because on occasion difference scores are treated as the absolute difference, where the direction of the difference is unimportant. Actually, they are not the same.

The first PROC is a study in simplicity, for it has only one classification variable and one dependent measure. When multiple dependent measures are stated, separate analyses are performed on each. MANOVA analyses must be requested as a special option. Obviously, the same design is applied to each analysis.

The next analysis requests two classification variables and then defines the model as having two main effects and one interaction. In this case, we have a completely crossed design, with four levels of the first effect and five levels of the second. In many ways, this is the most difficult aspect of using ANOVA.

The final PROC is very similar to the first except that the dependent variable is the positive difference between pre- and post-headaches. Since the CLASS and MODEL statements are the same, the first and third procedure steps could have been combined as follows:

```
PROC ANOVA;
CLASS EXCONDIT;
MODEL PSTACHE DIFACHE=EXCONDIT;
```

The same number of analyses would result. With complex designs, such combinations are more confusing than helpful.

A final comment—both CLASS and CLASSES are acceptable instructions and do not have to be grammatically matched to the situation.

Further advice on model writing is given in Section 8.6. The best advice for the beginner is to have the model statement checked by a statistical expert. Such an individual need not know anything about SAS because the form of model writing is fairly universal. SAS does, however, give an example of a strip-split-plot design. Social scientists probably do not need to learn this design immediately (if ever). It is handy for agronomy studies, however.

8.6   Pro Tips and Techniques for SAS

This section is aimed at anyone wishing to improve SAS skills. Although it may not be comprehensive, this section presents some of the author's experience with SAS. Certainly many useful techniques have been omitted, but this section should move the reader closer to pro status.

The following topics will be treated: abbreviations, plurals, and hidden conventions; order of statements in SAS; SAS syntax; data transformations and conversions; and SAS error messages. These topics are independent of each other and should be read as needed. The beginner, of course, should read them all.

8.6.1   Abbreviations, Plurals, and Hidden Conventions

A basic rule for SAS is do not abbreviate, although there are of course some exceptions. Variable names may be abbreviated to the shortest nonambiguous term. In the previous exercises, if AGE were the only variable that began with the letter A, then it could have been abbreviated to A. However, in many of the exercises the variable AVGVOTE was also calculated, thus requiring the abbreviations AG for AGE and AV for AVGVOTE. Variables such as VOTE76 and VOTE72 cannot be abbreviated in any way. Abbreviation of variable names is possible only after they have been named in full once. Abbreviation of a name in an INPUT statement would define the variable as the shortened version, and the full version would not be correct. Removal of vowels to form abbreviations is not allowed.

All SAS keywords must be spelled exactly as given. Certain keywords are in themselves abbreviations. Some of these can only be used in the shortened form; for others, it is possible to substitute the full word. PROC, VAR, and STD may be replaced by PROCEDURE, VARIABLE, and STANDARD, respectively. ID, OBS, FIRST OBS, GEN, and CORR may not be replaced by the longer versions of their names. There are more of these types of abbreviations in SAS, but they are usually obvious. (There are also some that are so esoteric that if you do not recognize them, then you probably would not want to use them.) Here is one last note on abbreviations. In many procedures, SAS gives the user a set of interchangeable keyword names, e.g., NCLUSTERS=, NCL=, N=. When you are reading SAS, watch for these interchangeable names, as they are sometimes hidden.

Plurals are unique problems in SAS. Some statements allow plurals, some do not, and others specifically allow either the singular or the plural form of a word. For example, CLASS or CLASSES is acceptable in the ANOVA procedure, but only CLASS is allowed in the descriminate analysis (DESCRIM). The best advice is to forget English grammar and to use the statements exactly as given.

There are three troublesome conventions in SAS; the first is hidden, and the other two are confusing. All three are undoubtedly present in SAS-79 and SAS-79.5, but they are not stated very explicitly. The hidden convention is the use of character constants in SAS statements. A common example is coding a variable T for True and F for False. Such values must be enclosed in apostrophes (single quotes) when used in instructions but not when used in the data. For example, we could have

```
NTRUE=0;
IF Q1='T' THEN NTRUE=NTRUE+1;
```

Exactly where the character value appears does not change the requirement. Conversely, apostrophes are not used in SAS TITLE statements. And in INPUT statements, variables are referred to by an appropriate character format.

The first confusing convention is the choice of options within any SAS procedure. There are in effect three common types of options: procedure options (PROC), optional statements, and options to procedure statements. PROC options are always part of the procedure step itself; that is, they always appear before the semicolon that ends the statement. The most common of these is the DATA= option, which allows a data set other than the most recent to be accessed by the procedure. Certain procedures may have several of these options. For example, we might write (NEWSET is a data set name chosen for this example):

```
PROC PLOT UNIFORM DATA=NEWSET;
```

Optional statements stand alone and may or may not be required. Each procedure is unique in this respect. ANOVA has eight such statements, and the subprogram PLOT has only one. The critical point is that if such statements are used, they must not be part of the PROC statement. The statements that define a particular analysis or data operation often have options, but, unlike the preceding options, they are set apart by a slash (/) even though they are not separate statements. For example, in the PLOT procedure we might have

```
PROC PLOT UNIFORM DATA=NEWSET;
BY SEX;
PLOT AGE*IQ/VZERO HZERO;
```

In this instance, VZERO and HZERO are options within a statement, UNIFORM is a PROC option, and BY is an optional statement.

The second confusing convention concerns what *must* be included in a given procedure to achieve correct functioning. In some descriptions, SAS tells the reader what is absolutely required; for example, the TTEST procedure must have a CLASS statement. On the other hand, the ANOVA procedure does not explicitly state that the CLASSES and the MODEL statements are necessary. There are hundreds of models possible and no logical means by which SAS can choose a particular type. So, although SAS does not say that these statements are a must, they are. Other procedures are similar to ANOVA. Of course, the nature of the statistics often makes certain aspects essential, thereby rendering it difficult for the individual who is just learning a new technique. The only workable solution is to trust your statistical common sense and this system's excellent error messages. If you overlook something essential, SAS will tell you about it (see Section 8.6.5).

8.6.2 Order of PCL Statements

Of the three analysis packages covered in this text, SAS has by far the most direct and simple structure for its control statements. As discussed previously, there are data sets and the instructions to establish, modify, and retrieve them, and there are procedures (programs) and the instructions to execute them, to choose options, and to direct output. The possible combinations of these two general steps are nearly limitless. A given analysis must begin with a data step or must retrieve a previously created data set, but it does not have to have any procedures (e.g., to establish a data file for later analysis). There might be 1 data step and 25 procedures, or there might be 25 data steps and only 1 procedure.

We see that no definite statement can be made about the form of a typical SAS analysis. The thing to keep in mind is the concept of a file. SAS treats each file as a whole unit, and usually works with one at a time. The exception is that SAS can merge several files into one. Thus, an analysis might involve a total of 18 files but work with only 1 at a time. SAS is not constrained to work on multiple files in any special order except as the logic of the problem dictates. For example, you could request all pertinent files from SAS and then analyze them by naming them in the PROC step, or else you could retrieve the files as needed for each analysis. One way to keep things straight is to think of SAS as a simpleton that can only work with what is currently in its hands. In complex problems, special effort must be devoted to remembering which file is the current file. Recall that if a file is not explicitly named, SAS will assume that the most recently established (current) file is to be used.

Any number of temporary files may be created without changing the original. The duration of a modification is a constant point of confusion. For example, if a character code is converted to a numeric one in a file called PRACTICE, then the modification exists only in that file. Consider the following:

```
DATA RAWDATA;
INPUT;
(insert input specifications here)
CARDS;
(insert original data here)
DATA NEWDATA1;
SET RAWDATA;
IF Q1='T' THEN Q1=1;
IF Q1='F' THEN Q1=2;
NEWQ1=INPUT (Q1,1.);
PROC (insert second analysis details here);
DATA NEWDATA2;
SET RAWDATA;
PROC (insert third analysis here);
DATA NEWDATA3;
SET NEWDATA1;
PROC (insert fourth analysis here);
```

The file or data set named NEWDATA3 would contain the conversion of Q1 to NEWQ1, but NEWDATA2 would not. Clearly, if a fourth file were created from NEWDATA3, it would also contain the conversion. Very long chains of file manipulations should be avoided; it is too difficult to determine the effects of complicated manipulations.

8.6.3   Syntax

A common difficulty with SAS syntax occurs with variable lists. There are two separate types—sequenced and unsequenced. Sequenced lists must end with a number or numbers, as in Q1 or SUBJECT34. This kind of list is indicated by a single minus sign, e.g., Q1—Q15. Unsequenced lists are indicated by two minus signs, as in IDN——Q50 or GENDER——AGE. SAS can correctly interpret these statements only if all variables from the start of the list to the end have been previously defined. This means that unsequenced lists may not be used in an

INPUT statement. After the variables have been defined, SAS does know which and how many variables are between GENDER and AGE.

A sequenced variable list may be used in an INPUT statement because SAS is able to calculate and provide the intervening variables. But no matter what type of list is used or when it is used, the implied variables must in fact exist. Precisely what will happen if this rule is violated depends on the type and characteristics of the format associated with each variable.

Implied lists of either type may be used throughout SAS. However, implied lists cause problems when used in complex IF and IF-THEN statements. When this situation occurs, SAS will simply tell you that xxx is not allowable.

When the value of a character variable is used in an SAS statement, it must be enclosed by apostrophes. This allows it to be distinguished from a regular variable name. Consider the following two lines:

```
IF SEX='F' THEN GROUP=1;
```

```
IF SEX=F THEN GROUP=1;
```

The second statement asks SAS to set the value of GROUP at 1 if the two variables SEX and F are equal, which is not a test for the character value F. Tests of the equality of variables are legitimate and useful; but if those are the tests that you want, be sure to use the single quotes.

Contrary to this usage of character variables, SAS does not require quotes for titles or labels. More critical for SAS is the effect of a missing semicolon at the end of an instruction. Consider the following lines:

```
VAR MEN WOMEN
TITLE OPINION DIFFERENCES;
```

In this circumstance, the error message from SAS will state that the variable TITLE is unknown. This is confusing when you know perfectly well that your variables are MEN and WOMEN. However, like SPSS and BMDP, SAS reads until it encounters an end-of-statement signal. Here, the only signal appears after DIFFERENCES. When syntax errors are involved, error messages can be quite misleading.

Another tip on syntax concerns the input convention of each signs (@@). Experience indicates that considerable confusion occurs here among individuals new to SAS. This convention provides SAS with a *free,* or *stream,* input option. The data are continuous in the sense that the observations (cases) form an uninterrupted stream as opposed to the usual pattern of completing an observation and starting a new line (card). Data in this form are very hard to proofread, so this practice is not advisable. List input is typically the format chosen for use with this convention. As an example, consider the following:

```
DATA PRACTICE;
INPUT IDN AGE IQ GPA @@;
CARDS;
```

If each of these variables were exactly three digits wide with one space between, then five groups of observations would fit on each line (card), and SAS would continue to read records until there were no more. Slashes are not required.

Another syntax issue that is troublesome to many is the SAS requirement that once a variable is formatted as character or numeric, it is permanent. A simple process such as obtaining the mean number of true and false answers is not possible unless the special conversion technique outlined in Section 8.6.4 is used. There is nothing wrong with this lack of interchangeability; it just appears incorrect to those researchers who are experienced with SPSS and BMDP.

Some clarification is also needed on the issue of the SAS use of the logic of arrays for manipulating a series of variables. When confronted with the necessity of modifying 20 variables all having different names and having a scattered order in the INPUT data set, one can write a series of 20 individual manipulations, or one can place the variables in an array, which SAS will then manipulate. Then each variable will be changed in accordance with the array manipulation. For example, if each of nine variables were coded with 'DNA' for a response of "does not apply," then the following test would save time:

```
ARRAY QUEST Q1 Q2 Q3 QA1 QA2 QA3 QB1 QB2 QB3;
INPUT IDN SEX AGE Q1 Q2 Q3 Q4 Q5 GPA IQ QA1 QA2 QA3 QA4 QA5 SATSCORE QB1 QB2
  QB3 QB4 QB5;
DO I=1 TO 9;
IF QUEST='DNA' THEN QUEST=.;
END;
```

Let us ignore the DO statement for the moment. The above statements will test the nine variables in QUEST for the presence of a 'DNA.' When a 'DNA' is found, the missing value code (the period) is substituted. This is certainly easier than conducting nine individual IF tests. When there are dozens of tests to be made, the savings in time, effort, and money are considerable. Note that not all variables are tested for a 'DNA' code.

The preceding statements also illustrate another of the more common areas of SAS syntax errors. Recalling prior programming experience, many users tend to write QUEST as QUEST(I), which is incorrect. SAS uses implied indexes, also called subscripts. This means that QUEST has the implied index I. For the statistical package user, there is no advantage to either form, but it is necessary to remember the correct one.

The preceding example also has what is called a DO loop. Such loops are a way of obtaining a repeated action. In the example, the index (I) is first set at 1, and succeeding statements are executed until the END is encountered. In this case, there is only one statement to execute. After the END statement, SAS loops back to the DO, advances the index to 2, and restarts the cycle. Such cycles may start at any number and may advance to any number by any increment. Compound loops are also possible. Many programming languages offer the DO loop. If arrays and loops appear to be an advantage, a good programming text will be useful, especially since the description of them in SAS is quite brief.

A final point to remember about SAS syntax is that errors tend to compound and the error messages from SAS become less and less useful. SAS can best help you on the first error. After that, use caution.

8.6.4   Data Transformations and Conversions

SAS contains certain very powerful provisions for data transformation and modification. Most conversions are accomplished through user-written programming techniques. A series of instructions must be written to perform the transformations. Although this option exists in all of the analysis packages discussed in this text, SAS requires more programming expertise than do SPSS and BMDP. One way to look at the issue of programming expertise is to note the presence of many convenient features in SPSS and in BMDP for making transformations. An excellent example is the RECODE facility in SPSS. With this, one may transform, group, and eliminate data values of any type. To recode a variable in SAS, particularly from an alphabetic code to a numeric one, one must make a special effort. Exercise 8 in Section 8.5 includes such a conversion. Here is the same conversion without the irrelevant material:

```
INPUT VOTE76 $1. +1 VOTE72 $1. +1 VOTE68 $1. +1 VOTE64 $1. +1 VOTE60 $1.;
ARRAY VOTE(I) VOTE76 VOTE72 VOTE68 VOTE64 VOTE60;
ARRAY NVOTE(I) NVOTE76 NVOTE72 NVOTE68 NVOTE64 NVOTE60;
DO I=1 TO 5;
IF VOTE='D' THEN NVOTE=1;
IF VOTE='R' THEN NVOTE=2;
END;
```

The contrast may be seen by noting that SPSS could do this conversion in one statement. This is not to say that SPSS is a better package than SAS; rather, it is a point of difference.

Analyzing this example step by step, we see that the INPUT statement is routine. The two ARRAY statements define groups of variables that may be referred to by an index, or subscript. (See Section 8.6.3.) The DO statement sets the indexing parameters; in this case, they are from 1 to 5 in increments of 1. The actual data conversion begins with the IF statements. Two IF statements are preferred over the IF-THEN-ELSE form because only R's and D's are tested. The ELSE form would make all other letters also equal to 2. With two IF statements, "bad" data are not converted inadvertently. The net effect of the two IF statements is to test each value of VOTE for 'D' or 'R' and to set NVOTE equal to 1 or 2, respectively. SAS carefully notes such conversions, stating precisely where they occur. However, SAS still considers VOTE76, and so on, to be character variables, each case containing R's and D's. Thus, any later PROC that is restricted to numeric variables will result in an error condition serious enough to suspend the analysis. The final ARRAY resolves this problem by defining a new variable (NVOTE) having the value of VOTE. Thus, NVOTE has the value of VOTE but is defined as a numeric variable.

A variable cannot be defined as numeric at input and alphabetic in the data. SAS will spot such conditions and reject them. These means of conversion may seem awkward but are the full equivalent of the RECODE facility. Also, it is not always necessary to use a DO loop. Consider the following:

```
DATA COLLEGE;
INPUT AGE 10 2.0 +1 GPA 3.2;
GROUP=0;
IF AGE LT 16;
IF AGE GT 65;
```

The data set named COLLEGE will contain a subset of the raw data, i.e., those who are less than 16 years of age and those who are greater (older) than 65. In order to form groups for analysis, we would have this:

```
DATA COLLEGE;
INPUT AGE @10 2.0 +1 GPA 3.2;
GROUP=0;
IF AGE LT 16 THEN GROUP=1;
IF AGE GT 65 THEN GROUP=2;
IF AGE GE 16 OR LE 65 GROUP=.;
```

These instructions would form three groups: over 65, under 16, and 16–65. The last group is considered missing. Clearly, very elaborate conversions and selections can be made using algebraic logic, logical operators, Boolean logic, DO loops, and arrays. The only trick is to keep the syntax correct to achieve what is desired.

Transformation of data values such as the square root of x and the square root of $x + 1$ is quite routine in SAS. For example, we could write these statements:

```
DATA THESIS;
INPUT IDN 2. +1 LATENCY 3.2;
LATENCY=SQRT(LATENCY+1);
CARDS;
```

In this instance, the data set THESIS would not contain the original data; instead, it would contain the square root of each data value plus 1. A more careful procedure would be to retain the original variables and then to name a new variable for the conversion or else to create a new data set from the original for the converted data. Either way, the original data would be available.

SAS lists a large number of fixed functions, all of which work in the manner described above. Remember that algebraic functions have explicit effects on the data, and it is possible to find oneself searching for data that no longer exist.

SAS provides two additional features that are useful in data manipulations: DELETE and DROP. There are both a DELETE statement and a DELETE procedure. The procedure is used for eliminating entire data sets from SAS files, and the statement is used for eliminating specific observations from a data set. It is the DELETE statement that is of concern here. It works only when a data set is being created; for example:

```
DATA COLLEGE;
INPUT AGE 10 2.0 +1 GPA 3.2;
IF AGE GT 15 AND LT 66 THEN DELETE;
CARDS;
```

These instructions accomplish the same selection as the multiple IF instructions discussed earlier. Clearly, this statement could be used in a variety of ways, but it applies only to observations.

The DROP statement applies to variables and allows certain variables to be excluded from a data set. This is very useful when the amount of data are excessive or when data, once they have been processed, are to be removed. Assume that you have MMPI (*Minnesota Multiphasic Personality Inventory*) data on 10,000 individuals and that you are only concerned with the total number of true answers for each individual. This mass of data could be considerably reduced as follows (remember the implied index in SAS DO loops):

```
DATA MMPITEST;
ARRAY Q(I) Q1-Q560;
INPUT (insert format for 560 questions here);
NTRUE=0;
```

```
DO I=1 TO 560;
IF Q=1 THEN NTRUE=NTRUE+1;
DROP Q1--Q560;
END;
CARDS;
```

The data are thereby reduced from 5,600,000 individual numbers to an even 10,000. (A KEEP instruction that is the opposite of DROP is also available.) To summarize, one can DELETE observations, DROP variables, and DELETE (in a PROC step) data sets.

There is one final point to be made about DO loops in SAS. The ARRAY instruction and the DO instructions must be stated within the same data step. For example, using the following in the MMPI example would not work:

```
DATA MMPITEST;
ARRAY Q(I) Q1-Q560;
INPUT (insert format for 560 questions here);
DATA MMPIRUNI;
SET MMPITEST;
NTRUE=0;
DO I=1 TO 560;
IF Q=1 THEN NTRUE=NTRUE+1;
DROP Q1--Q560;
END;
CARD;
```

To function correctly, we must have the ARRAY Q within the second data step as follows:

```
DATA MMPITEST;
INPUT Q(I) Q1-Q560;
DATA MMPIRUNI;
SET MMPITEST;
ARRAY Q(I) Q1-Q560;
NTRUE=0;
DO I=1 TO 560;
(and so on)
```

All things considered, SAS is extremely flexible in its allowable data manipulations, even though a certain level of skill is required. The user must guard against conversions that result in logical traps or inaccuracies. Always make provisions for detecting data that are outside of the known parameters; otherwise, you could end up analyzing errors instead of data.

8.6.5   Error Messages

Of the three major packages, SAS provides the clearest and most helpful error messages. They locate the trouble spot precisely. Locating an error does not always guarantee that the problem can be immediately resolved, but SAS gives the most assistance possible with its wide variety of specific messages.

As was pointed out in the syntax discussion, omission of a semicolon from an SAS statement can cause unpredictable effects. SAS will usually interpret the next statement as part of the statement without the semicolon. Unless the error is in an optional statement, processing will be suspended. It is this error that can make a disaster of the use of variable names that are similar to SAS names. For example, if you had a variable named PROCONE, SAS might print an error message about problems in the PROC statement if the error was an omitted semicolon. Therefore, avoid SAS keywords when you are assigning variable names.

While many errors will cause a suspension of processing, it is frequently only the current step that is ended. If possible, SAS will continue processing the subsequent steps. Occasionally when this happens, the following message will result: NO OBSERVATIONS IN DATA SET. This indicates that the error was in the data step and caused all of the data to be missed. The default that suspends processing when errors are encountered can be suppressed (see the manual for SAS-79, Appendix 2). Do this suppression with caution, as the results are unpredictable.

The SAS beginner should take error messages with a grain of salt because they can be misleading. (When a message is received that explicitly diagnoses a problem, it should be considered as good fortune.) It is even more important to exercise caution when SAS prints multiple error messages. Subsequent messages are often unreliable because they are built upon the first error and its message. It is often next to impossible to determine the logical process that led SAS to a specific message after it has printed the first error message.

When using SAS with a terminal system, it is often possible to recover from an error because each error is translated as it is submitted. One simply restates the instruction or deletes the most recent step and begins again.

8.6.6 Special Formats

In previous sections, the standard formats of SAS were discussed. This section takes up the special formats of SAS, for which it is unique among the three major packages. The special formats vary from Roman-numeral and social-security-number types to 13 different variations of date and time values. Only those special formats that are potentially useful to the social science researcher will be taken up here.

The BZ$w.d$ format is useful for data that were originally prepared for SPSS or BMDP. It defines blanks as zeros, which SAS does not normally do. When using this format, remember that having blanks represent zeros applies only to the variables that use this format. This procedure is used just as if it were the standard numeric format *w.d* but with BZ as a prefix. There is no BZ*w.* format.

Some of the SAS special formats are for output only, some are for input only, and others are for both input and output. For example, ROMAN*w.* is for outputting integers as Roman numerals—decimals not allowed. Many of these formats are strictly for the data processing professional and can be ignored by most social scientists.

SAS offers four different character formats. The format $*w.* has already been discussed. The $CHAR*w.* format allows blanks to be used as legitimate characters. If your data contain blanks that are to be included, there is no choice but to use this format. The other character formats are for the data processing expert.

Of the thirteen different date and time formats, the three important ones are date*w.*, mmddyy*w.*, and yymmdd*w.* (If you have not guessed, m is for month, d is for day, and y is for year.) Some confusion exists because the first of these formats (date*w.*) requires that the data be punched in military form, i.e., day/month/year, with the month spelled out in a three-character abbreviation. 27JUN84 and 08SEP60 are two examples of this style. Blanks are not allowed, and the period must be included. Confusion arises because the SAS manual states that date*w.* has the form ddmmyy, which can easily be mistaken for the names of the other two date formats. The second format is mmddyy*w.* and requires that the date be stated in month/day/year form. The third format (yymmdd*w.*) is similar but uses the opposite order. Consider the following example:

```
DATA STUDYONE;
INPUT IDN 3. +1 TESTI DATE7. +1 TESTII MMDDYY6. +1 TESTIII +1 YYMMDD6.;
CARDS;
001 05JAN84 010684 840107
```

From this INPUT statement, SAS would expect: the IDN number to be three digits wide; the variable TESTI to be seven digits wide and in the DDMMYY form; a third variable, TESTII, to be six digits wide and in the form MMDDYY; and a fourth variable, TESTIII, to be six digits wide and in the YYMMDD form. From the first data card, we can see that the individual whose IDN was 001 was tested on consecutive days in January, 1984. In all of these instances, *w.* specified the width of the date. Wider specifications can be made when necessary. All SAS date values must be justified on the right if they are in fields that are wider than needed. Be sure to read the SAS manual on dates because there are several variations not mentioned here.

COMPUTER USAGE FOR SOCIAL SCIENTISTS

It is important to realize that SAS does not work with dates directly; rather, it translates them to a value representing the number of days from January 1, 1960. In this way, dates may be used as valid numbers without further translations. Remember that after any manipulation of date values they may possibly no longer represent a date. When working with SPSS and BMDP, the user is advised to always state dates in the yymmdd form so that they approximate a rational number system. With SAS, this is not necessary because of the automatic translation feature.

Other, more exotic forms of manipulation of dates are also found in SAS but are beyond the scope of this text. Minutes and seconds as well as time of day are also available through the format process. The update manual for SAS-79.5 provides even more formats.

8.6.7 Special SAS Functions

Some functions such as SQRT have already been discussed. This section will present a brief comment on SAS time and date functions. Note that these are *functions*—not formats. For example, if you wished to use the current time of day as a value, you could write this:

```
DATA;
SEED=TIME(    );
```

The variable SEED would then contain the time of day at which SAS encounters the instruction. Such a value could be used to provide a starting seed for Monte Carlo methods. Take care not to confuse this function with the format of the same name. Indeed, the explanation in SAS for the TIME function is somewhat misleading because it contains the TIME format as well.

The other principal use for these functions is to produce output labels; e.g., analyses and data sets may be labeled in seconds, hours, Julian calendar dates, etc. These are useful procedures for those who are producing reports or graphs that require this form of labeling. Such procedures are not necessary in formal analyses because SAS will always tell you when the analysis was performed.

8.7   Pro Tips and Techniques for Specific Procedures

SAS offers 36 individual statistical procedures and 17 utility operations. Rather than attempting to discuss each, this section will treat groups of related procedures and will present only those areas where confusion or trouble may occur.

8.7.1   The CORR (Correlation), FREQ (Frequency), and RANK Procedures

The correlation (CORR) procedure is a very routine SAS analysis with only two tricky areas. The first is the way in which different types of correlations are requested. The correlations are obtained as options to the PROC CORR statement. The assumption is that we will receive PEARSON correlations, in the absence of a specific request. However, the option PEARSON must be stated if you wish both PEARSON and KENDALL (or SPEARMAN) correlations. The second difficulty is in outputting matrices. If we simply request a covariance (COV) matrix, it is not automatically put into the output file. An individual request must be made, although the matrix would be printed. This situation arises when matrices are needed as input to other procedures.

The frequency (FREQ) procedure can produce one-way, two-way, and multiway tables for numeric, character, and combinations of numeric and character variables. Numeric and character variables will have an ascending order, with character variables limited to the first 16 characters. Some caution is required when using character variables due to the oddities of the sorting sequence. All names with only initial capital letters precede any names of all capital letters; e.g., Ziegler would precede ADAMS. See the SORT procedure for the full sequence. With numeric variables, the sequence of categories is always what one expects (1, 2, 3, 4, etc.), but character values will produce some surprises.

A related problem is the number of groups that FREQ determines. In many cases, steps must be taken to restrict the number of groups formed. For example, a researcher interested in cyclic variations in animal behavior might allow rats to run an activity wheel 24 hours a day. If measured to the minute, the time variable would have 1440 (60 × 24) real categories. Since this is a four-digit number, SAS would be prepared to form 9999 categories. The SAS manual suggests using the FORMAT procedure to restrict the number of groups. Another way to accomplish this limitation would be to create a new data set with data manipulation statements to restrict the values of variables. Unless you can afford thousands of pages of output, think about this problem before turning SAS loose. The SAS manual points out that multiway tables are another source of runaway output.

The RANK procedure does just what its name implies; it computes ranks for a numeric variable and creates a new data set containing the ranks in place of the originals. The options are just what would be expected from a ranking program: ties, ascending or descending groups, and percentages. Several variations on computing normal scores from the ranks are included. Be careful to observe the assumptions of the

options; for example, PERCENT assumes the HIGH method of re-
solving ties.

8.7.2   The MEANS, SUMMARY, and UNIVARIATE Procedures

The MEANS procedure produces much more than just means. In
addition to arithmetic means, MEANS produces a large selection of
descriptive statistics that range from the number of observations to
kurtosis. The variable to be used must be named, and variables must
be moved to an output file. In addition, each statistic desired must be
indicated in the OUTPUT instructions. The procedure does not automatically put all calculations in the output file—only those that are
specifically named. When the calculations are printed, the procedure
always produces a subset of the variable statistics. For example, if
skewness is desired, it must be requested in the PROC instruction. The
PROC MEANS instruction treats missing values in a way that occasionally produces surprising results. When we carefully think through the
effect of missing values, it usually explains why certain statistics were
not calculated. Note that this procedure may also be used to obtain a
paired t-test. (See the discussion in the SAS manual under TTEST
procedure.)

The SUMMARY procedure is very similar to MEANS but cannot
produce a printed output. Its purpose is to pass calculations to another
procedure, which could of course be PRINT, or the printing utility.
The other major feature of SUMMARY is the ability to form subgroups
without using the SORT procedure and BY option. Output control is
critical to this procedure and is achieved through OPTIONS instructions. In this case, an output option is the name of a statistic, e.g.,
RANGE. It is strongly recommended that you always rename the
variables. If you do not, even a little inattention can result in several
variables having the same name. Moreover, while this duplication is acceptable to SUMMARY, other procedures will not accept this condition.

The UNIVARIATE procedure is similar to other descriptive programs,
but it provides a more detailed picture of the variable. It operates in
much the same way as the SUMMARY and MEANS procedures; the
user chooses options, names variables, selects statistics for output, and
renames output variables. If only printed output is desired, no selection
of statistics is necessary, as they will automatically be printed.

8.7.3   The ANOVA, DUNCAN, FUNCAT (Functions of Categorical Responses), and GLM (General Linear Models) Procedures

For the beginner, ANOVA is the procedure of choice for analyzing experimental statistical models. GLM (General Linear Models), discussed
later in this section, is an alternative method that places a far greater

demand on SAS resources and requires greater expertise on the part of the user. However, ANOVA is for balanced designs only, that is, designs with equal numbers of observations in each cell (equal n's). There are exceptions to this that are mentioned in the SAS manual. As the manual points out, it is up to the user to meet these requirements. ANOVA does not make any checks on the data.

The heart of PROC ANOVA is the MODEL statement. Again, a wide variety of models is available, and it is not possible to check for their correctness. Crossed factors are indicated by the asterisk (*), and nesting is shown by parentheses. The variable outside of the parentheses is nested within the variable named in the parentheses. Thus, sex within age would be written SEX(AGE). A variable may be nested within two or more variables. Crossed and nested notations can also be combined. Beginners often encounter difficulty in writing correct models. This is not an effect of using SAS and ANOVA; rather, it is the result of statistical inexperience. The ANOVA procedure offers little help in this regard; the user must know the statistics well enough to write correct models.

A CLASS or CLASSES statement must also be written to define the dimensions of the analysis. Classifications are known by many names that depend on the preference of the statistician who determines the notation and the verbal description of a particular statistic. Factors and categorical variables are two common names. The point is that the CLASSES statement names variables and the MODEL statement describes their relationships. The SAS manual inadvertently implies that only single-character names are allowed in CLASSES and MODEL statements. This is not so; any name may be used as long as it conforms to SAS variable name conventions. Finally, in writing ANOVA models, the order of effects is main effects, crossed effects, and then nested effects. Of course, a given model might not have one of those types of effects.

If Duncan (or Duncan-Waller) tests are desired, MEANS must be requested. These statistics are part of the MEANS statement and may appear only as a MEANS option.

Beginners should avoid the ABSORB statement, as it is a somewhat exotic technique. (Obviously, use it if you know how.) Likewise, the beginner should exercise caution when specifying error terms for specific F tests. This is a powerful option, but it is also a rather sophisticated one.

The MANOVA option is very useful but performs analyses in specific ways with which the user must be familiar. The SAS manual gives several references to these methods. MANOVA, without options, is used to obtain a special missing-values treatment; i.e., if any variable is missing from an observation, it is excluded from the analysis. This will not produce a multivariate analysis.

The DUNCAN procedure provides a means to obtain the Duncan's multiple range test and the Waller-Duncan K-ratio tests independently of the ANOVA and GLM procedures. As such, DUNCAN is very routine. However, it must be supplied with the requisite information. The user must provide the degrees of freedom (df), mean squares (MS), and F values (Waller-Duncan only) for each variable. These values can be provided by the user or obtained from any of several other procedures.

The functions of categorical responses (FUNCAT) procedure is strictly for experts. This is due more to its rare use than to its conceptual difficulty.

GLM is, to coin a phrase, the "big daddy" of SAS procedures. Not only can it do analyses of variance including unequal n's, but it can also handle covariance, several types of regression, and partial correlation. It provides extra features for each of these techniques. Also, there are techniques for confounded effects, estimates of missing values, least-square means for cells with unequal cells, contrast effects (no known limits), DUNCAN and other options found in the ANOVA procedure, the MANOVA option, two types of confidence limits, weighting, and residuals. Only the analysis of variance aspects will be discussed here. For other aspects, see Sections 8.7.5 and 8.7.6.

As a statistical approach to analysis of variance, GLM is very powerful. Indeed, there are some ANOVA models that can be analyzed by no other technique. However, this places a premium on the statistical knowledge of the user. Beginners should proceed with caution with GLM.

As in ANOVA, the MANOVA option produces the MANOVA treatment of missing values. The most difficult part of GLM is keeping the various types of models separate. If you cannot tell an analysis-of-variance model from a regression model, you will be working under very trying conditions.

The average SAS user should be very careful when using GLM for unbalanced and incomplete designs. Models written in a notation such as Y=A contain certain ambiguities. The interpretation of a specific part of the equation will depend on what other parts are present. This may become quite involved and very subtle. GLM does constitute a strong approach to these designs, but proceed with caution.

8.7.4   The NESTED, NPAR1WAY (Nonparametric One Way Test), PROHIBIT, TTEST, and VARCOMP (Computation of Variance) Procedures

NESTED is another analysis-of-variance procedure that can be done through GLM. It is efficient and less likely to cause errors for the beginner. All effects are assumed to be random and must be sorted by the classification variables. The usual procedure is to run the SORT procedure first and then to pass the data set to the NESTED procedure. The sorting must be in the order in which the variables are given in the CLASSES statement. Further, the nesting is specific; the second set is nested in the first, the third set in the second, etc. Using the notation for models in the ANOVA and GLM procedures, we would have:

```
CLASSES A B C D;
MODEL Y = A B(A) C(AB) D(ABC);
```

This does not mean that there is a MODEL statement in NESTED. Other forms of nesting cannot be analyzed by this procedure. With the exception of these points, NESTED is fairly routine.

The NPAR1WAY procedure performs a variety of single variable non-parametric analyses of variance. A CLASS statement is necessary even though there is only one variable. Further, the levels of the CLASS variable determine in part which statistics will be calculated. The user may select the desired tests, including a standard parametric one-way test. The only trick with NPAR1WAY is in selecting the most appropriate statistic from the many that are automatically generated. Unless you know precisely what you want, you will be required to study nonparametric texts.

The PROHIBIT procedure is usually used for biological assay data. Naturally enough, the user must state the variables of dose, number of subjects, and subjects responding to the dose in a variables (VAR) statement. From this brief description, it is obvious that this is another use-it-if-you-know-it procedure. Note that biological assay is not the only use of the general analysis technique of PROHIBIT.

VARCOMP (computation of variance), like PROHIBIT, is an advanced technique. It calculates variance estimates for most linear models composed of main effects, interactions (crossed effects), and nested effects. Continuous variables are not allowed as independent variables, but effects may be fixed or random. Three different approaches are offered, all of which are definitely not for the beginner.

The TTEST procedure is fairly routine except for three points. First, the name of the procedure is TTEST—not T-TEST, as in SPSS. Second, it is necessary to insert a CLASS statement naming a variable that has two and only two values. Other statistical packages will allow more than two test groups and will perform all possible tests between them. Often, such multiple t-testing is not considered to be a correct statistical design. SAS simply disallows it. Of course, multiple t-tests can be obtained with multiple PROC steps if required. Third, TTEST is not intended as a paired or correlated test. The SAS manual does document a way to obtain these tests through the MEANS procedure.

8.7.5   The STEPWISE, NLIN (Nonlinear Regression), SYSREG (Systems Regression), RSQUARE (Regression with R^2), and GLM Procedures

Of all the regression procedures in SAS, the easiest to use is STEPWISE. It is the recommended starting point for those users who are new to regression or to SAS. This procedure requires less advanced statistical knowledge and has a fairly typical SAS structure. The most difficult aspect of STEPWISE is choosing the exact form of stepwise regression. Careful reading of the SAS description and a good statistics text are all that can be suggested here. Options compound this problem. On the other hand, the experienced user will realize that the many options and types of regression make STEPWISE extremely versatile. Remember that any regression procedure can be costly when many dependent and independent variables are involved. Keep the problem manageable.

The NLIN (nonlinear regression) procedure is not for the statistically inexperienced. It requires that the user be able to provide partial derivations of the model for each parameter. If you can cope with the math, NLIN is a typical SAS procedure with no surprises. It does call for more mandatory statements than usual, but they are precisely what is required for nonlinear regression.

The SYSREG (systems regression) procedure is another powerful but advanced technique for regression situations in which the variables are jointly dependent (interdependent). The approach is through systems of simultaneous equations. An example of such a problem that might be of interest to a social scientist can be found in the topic of authoritarianism. A researcher might attempt to construct a regression model that postulates the theory that authoritarianism is predictable from the factors of rigidity, intelligence, and random error. And if rigidity is simultaneously related to authoritarianism, cultural norms, and random error, then both equations operate together. As an SAS procedure, SYSREG is typically complex.

The RSQUARE (regression with R^2) procedure is very simple; the only fixed requirement is a statement of the regression model(s). RSQUARE is used in the evaluation of many models by determining the percentage of the total variance captured by each model. RSQUARE is usually part of a general regression approach to a particular problem area and is not a stand-alone, isolated analysis.

As a regression procedure, GLM can do almost everything. Therefore, it is very general in approach, and the user must know precisely what is needed. There is little offered in the way of help for the inexperienced. For example, regression models and analysis-of-variance models occasionally appear to be similar. The distinction is that regression analyses never include CLASSES statements and analyses of variance always do.

As a procedure for those who are statistically sophisticated, GLM provides a wide range of options and special variations. Potential users of this procedure should always keep in mind that GLM uses matrix manipulation methods as its mathematical technique. (See Section 8.7.6 for more on GLM.)

8.7.6   The CANCORR (Canonical Correlation), CLUSTER, DISCRIM (Discriminant), FACTOR, GLM, GUTTMAN, NEIGHBOR, and SCORE Procedures

The procedures CANCORR, CLUSTER, DISCRIM, FACTOR, GLM, GUTTMAN, NEIGHBOR, and SCORE form a group of interrelated techniques that are used to determine the structural relationships among variables. Often, the variables are test items or survey questions. These techniques are rather advanced statistical topics and are beyond the scope of this text. Even the user who is familiar with the techniques should practice with simpler SAS procedures first. Several of these procedures depend on other procedures for preliminary calculations. This means multiple PROC steps and the use of OUTPUT statements. Be sure to give this output a name that is related to but different from the original. Also, remember that such output matrices can be very large. As a general rule, these matrices should always be stored on disk or tape and not on punched cards.

8.7.7   The AUTOREG (Auto Regression), SPECTRA (Spectral Analysis), MATRIX, and PLAN Procedures

The procedures AUTOREG, SPECTRA, MATRIX, and PLAN are also very specialized statistical tools. Of special note is MATRIX. While not truly a statistical procedure, it can be used to do almost any analysis imaginable—as long as one knows matrix algebra sufficiently well. It is, along with GLM, one of the most powerful SAS procedures. It could conceivably be used in conjunction with a course in matrix techniques.

AUTOREG (auto regression), SPECTRA (spectral analysis), and PLAN are procedures which are far beyond the purposes of this text. They are structurally quite typical, however, of SAS.

8.7.8   The CHART and PLOT Procedures

The CHART procedure is a flexible approach to the creation of a variety of bar graphs as well as other types of charts. Raw data, sums, and means can be plotted by request. User-defined variables may also be graphed if they have been created in a preceding data step. There are several points to keep in mind when using CHART. The dimensions of the graphs are automatically determined unless AXIS statements are written. Continuous variables will often produce charts with peculiar dimensions if LEVELS has not been specified. For example, a variable such as age would have a bar for every age from zero to the oldest person's age; coding LEVELS=10 would reduce the number of bars to ten, making a much more readable graph. The DISCRETE statement should likewise be used to eliminate the assumption in CHART that all numeric variables are continuous. When gender is coded 1 and 2, CHART will make provisions for genders coded 1.3, for example, unless DISCRETE is used.

The PLOT procedure is used to graph two variables on the same grid; these are often referred to as bivariate plots. This procedure provides a host of features, including superimposed plots and contours. Like CHART, PLOT scales the X and Y axes automatically. Both axes may be defined by the user with the YAXIS and XAXIS statements. It should be remembered that interpolation between points is linear. To obtain an accurate nonlinear axis, the user must calculate and state every increment. A useful procedure is to plot the values of a third variable at the points representing the X and Y variables. When using CONTOUR, remember that the resulting plot will be rather unsatisfactory unless there are many data values. In some situations, values may be generated to fill the contours. When plotting equations, you must calculate the X and Y values in the data step and not in the PROC step calling PLOT.

8.7.9   The FASTCLUS (Fast Clustering), PRINCOMP (Principal Components), VARCLUS (Variable Clustering), and STEPDISC (Stepwise Discriminant Analysis) Procedures

The procedures FASTCLUS (fast clustering), PRINCOMP (principal components), VARCLUS (variable clustering), and STEPDISC (stepwise discriminant analysis) are new additions to SAS and are available only in installations running SAS-79.5. They add significantly to the statistical abilities of SAS but are aimed at the experienced user. Learn what an eigenvalue is before tackling these programs. The math is not impossible—just uncompromising.

8.7.10   The RSREG (Response Surface Regression) and REG (Regression) Procedures

The RSREG (response surface regression) procedure (added in SAS Version 79.5) is strictly an advanced technique and is not the place to learn SAS. The REG (regression) procedure (also in Version 79.5) is, on the other hand, a reasonable place to learn SAS and regression techniques. The novice should let SAS determine the specifications initially and then begin using the options after gaining some experience. The real danger in REG is the myriad of options. Do not let them intimidate you. Indeed, one virtue of REG is that you will not soon outgrow it.

8.7.11   The TRANSPOSE and PRINT Procedures

Available in SAS-79.5, the TRANSPOSE procedure changes rows to columns, and vice versa. If you have created a data set with variable-by-subject form instead of the usual subject-by-variable form, this procedure may save the day. However, transposition of rows and columns is not without certain ambiguities and difficulties. TRANSPOSE might well fail. The best advice is to avoid the need to use it.

The PRINT procedure expands the report-writing ability of SAS. It allows extensive custom tailoring of output. It is very straightforward and can be used by individuals at all levels of skill.

8.7.12   The BMDP Procedure

The BMDP procedure is a very useful SAS procedure in that it allows SAS to pass a data set to BMDP for analysis. The exact details will of course depend on which version of the BMDP system is available. The user creates the usual SAS data set and then in a PROC BMDP step states which of the BMDP programs is desired. Such file-swapping between analysis systems requires that the UNIT and CODE statements be used in BMDP itself. It is not very clear from the description of the PROC step to call BMDP that the user must, in addition to SAS statements, write the statements that would be necessary if BMDP were used directly. The typical BMDP paragraphs and sentences are included as part of the BMDP procedure. The two types of instructions are kept separate by a special SAS statement (PARMCARDS), which precedes the BMDP instructions. BMDP can likewise request an SAS analysis. However, the potential user must understand the other system.

8.7.13   SAS Utilities and OS Utilities

OS utilities in this context are IBM operating system utilities, and as such they will vary among installations. In general, most of these procedures are directed at data-base management and are only of incidental concern to most social scientists.

The SORT and STANDARD procedures are of the most interest. SORT has been discussed above, being the simplest of the SAS procedures. The only trick is to remember that SORT uses all of the numbers, symbols, and letters in the sort sequence. Thus, ADAM197 appears before ADAMS. Experience is the best teacher here.

The STANDARD procedure provides a means to convert data to standard scores. The mean and the standard deviation may be set to any desired value; for example, some IQ tests have a mean of 100 and a standard deviation of 16. This is a most useful capability when scores on different measurement scales must be compared. The measurement itself must be interval or ratio.

The CONVERT procedure allows SAS users to access files that are part of several other analysis systems, thereby making them SAS files. This facility works only from system files and not from raw data files. If there is no local option for saving files in BMDP or in SPSS, then it does not matter whether or not SAS is available; CONVERT will not work. Each of these conversions is slightly different, so pay close attention to the description in the SAS manual. Interconnection of analysis systems in this way presents an interesting situation. The SAS user can obtain a BMDP analysis, and vice versa; but this is not true for the SPSS user. SAS can read SPSS system files with the CONVERT procedure, and thus the SPSS user can obtain a BMDP analysis. Again, the opposite is not true. The SPSS user can obtain BMDP and SAS analyses, while BMDP and SAS users cannot obtain SPSS analyses. (*Note:* All of this interchange is predicated on the use of IBM systems. When different computers are involved, such interchange—while not impossible—would be a customized operation.)

$$===================================== \overset{\displaystyle 9}{=} =======$$

BMDP: Versions 75, 77, 79, and 81

It may sound a bit presumptuous to assert that this text can introduce BMDP better than the BMDP manual itself. Nonetheless, experience indicates that many students have great difficulty plowing through the BMDP manual alone. While this chapter is not truly an introduction, it will be a guide, or map, of the territory. The reader will need to simultaneously study the BMDP manual and this text.

The beginner should read this text through Section 9.2.1 and then proceed to the BMDP manual. After reading the introductory chapters, the novice can return to this book to complete Section 9.2. At that point, the user should be in a solid position to start running analyses. Section 9.3 provides suggestions and exercise problems on each of the major analysis areas that a beginner should consider, and it should be read in conjunction with the appropriate section of the BMDP manual. It is strongly recommended that novices start with program P4D. Section 9.4 ("Pro Tips and Techniques") will be of interest after some experience is gained with the package. Advanced users will probably find this the most helpful section.

9.1 Components Common to All BMDP Programs

The BMDP analysis package uses an instruction system with some similarities to the English language. The program control language

249

(PCL) uses English words and abbreviations in a paragraph-and-sentence arrangement. English grammar does not apply. The PCL statements only superficially resemble English, since many algebraic symbols are mixed with conventional words. These symbols are the same as those used in the FORTRAN programming language, as BMDP is predominantly a FORTRAN-based system. For example, the words IS and ARE and the algebraic equals sign (=) are all equivalent. What BMDP calls cases, a psychologist would call subjects, a psychometrician would call items, and a survey-oriented political scientist would call respondents. We can see in the following list that English grammar is *not* followed.

CASES IS 20.

CASES ARE 50.

CASES = 19.

CASES ARE 100.

From these four examples, we can discern several aspects of BMDP:

1. CASE is always wrong; the correct instruction is CASES.
2. Conventional English grammar does not apply.
3. Each sentence must end with a period; thus, there are four sentences above.
4. Blanks must be used to separate components whenever there is a possibility of confusion. BMDP will not understand the instruction CASESIS but will correctly interpret CASES=20.

Sentences are arranged into paragraphs as in English, but paragraphs are not indicated in the same way. The slash (/) is used to denote a paragraph. In versions published prior to August, 1977, the BMDP manuals instruct the user to end a paragraph with a slash. After August, 1977 (BMDP-77, -79, and -81), manuals say to start with a slash. Always use the newer convention. It is easier and less ambiguous. In terms of programming, the two conventions are equivalent, so the new procedure works with the older versions. Use it. It should be apparent that several sentences make up a BMDP paragraph. Many students seem to think that each sentence should begin with a slash, when in fact only each *new* paragraph begins with a slash. Furthermore, students often become very rigid about the order of sentences within a paragraph. Actually, the order of the sentences does not affect the instruction process, but sentences that are out of order can cause problems for the error-message system that is part of BMDP. See Section 9.4.2 for a complete explanation.

Another confusing point for beginners is the implication that each sentence must appear on a separate terminal line or punch card. This is not so. Consider that you have done a modest survey of the political conservatism/liberalism of a group of students. Assume that you have asked 20 research questions and 10 biographical questions, for a total of 30 questions. Considering each question to be a variable, we have 30 variables. Most BMDP programs require that you state the number of variables in this way:

VARIABLES ARE 30.

And if you have surveyed 98 students, you can use either of the following methods:

VARIABLES ARE 30. CASES ARE 98.

VARIABLES ARE 30.
CASES ARE 98.

In place of the word ARE, we could use IS or =. In actuality, BMDP does not read sentences; it scans from left to right until it encounters a period, then the next period, and so on. Sentences usually do not have to begin in a specific column. They may use the entire 80 columns of a terminal line or punch card (except, of course, when certain columns are occupied by something else, such as a paragraph slash). When a sentence or series of sentences exceeds 80 columns, you can proceed to the next line or card without doing anything special. Be very clear about this because you can get more bad advice on this topic than on any other. For the moment, let us consider sentence continuations. We will return to the bad-advice part later.

Continuation can be illustrated with another BMDP sentence—the NAME sentence. This is not a mandatory part of BMDP, but if you would rather have the results referred to by name instead of by a sequence number, this is the way to do it. In the political survey example, if nine biographical items were the first nine items on the data records and if they included age, sex, home town, height, scholarship, family income, whether the respondent lives on campus or commutes, number of siblings, and birth order, we could name the items in the following sentence:

NAMES ARE AGE,SEX,HOMETOWN,HEIGHT,SCHOLAR,FINCOME,RESIDENC,SIBLINGS,BIRTHORD.

Note that hyphens are not used as they are in a manuscript. Nor are there special characters anywhere on the first or second record that indicate continuation.

A typical arrangement that uses CASES, NAMES, and VARIABLES is as follows:

```
CASES ARE 98. VARIABLES = 30. NAMES ARE AGE,SEX,HOMETOWN,HEIGHT,SCHOLAR,
FINCOME,RESIDENC,SIBLINGS,BIRTHORD.
```

This example illustrates three important things about BMDP:

1. Continuations do not require special treatment and are therefore very easily accomplished. However, the well-meaning computer jocks found in every computer center are sure to tell you that your job did not run because you needed a slash in column 6 of a continuation. This is bunk! Something else is wrong. The slash is a FORTRAN continuation technique, and you are not writing in FORTRAN. It does not matter that BMDP is itself written primarily in FORTRAN. When you encounter this situation, say thank you, and then find someone who knows the BMDP system.

2. Alterations to sentences for separate analyses are more difficult to make when sentences are not on individual lines or cards. The advantage to punching one sentence per record (including its continuations) is the ease in modifying sentences for future analyses. In the above example, if we were to collect more data and have 305 cases, we would have to either create all new instructions or else, if CASES is on a separate card, merely change the 98 to 305. If there were a means of gaining an extra column in the middle of a card, there would be little difference between the two procedures. The general recommendation is to keep each sentence on a separate card. Those researchers who are using terminals instead of cards can use the insert facility to their advantage in this situation, but insertions greater than 80 columns will create minor complications that are well worth avoiding. Novices would be wise to keep instructions on separate lines.

3. There are rules and conventions to be followed when assigning names. Naming conventions in BMDP are fairly straightforward and cause problems only when violated. They are as follows:
   a. Names must be eight characters or less.
   b. All names must begin with a letter.
   c. Names after the first letter may be comprised of any letters or numbers.
   d. Names cannot contain any special characters, including blanks.
   e. Do not use a PCL (program control language) or SCL (system control language) operator as a name; i.e., do not call a variable JOB, VARIABLE, CASE, etc.

     f. Special characters and blanks may be used if the name is en-
       closed in apostrophes, e.g., 'F-INCOM' or '1ST-TEST.'

Students do not have many problems with names *per se,* but they do
have trouble with the *unnamed* variables and with the correspondence
among the order of the variables, the order of the names, and the order
of the actual data. All three orders must correspond exactly. Consider
the following:

data	001	18	1	1	2	9	7	3	,1
variable	$X(1)$	$X(2)$	$X(3)$	$X(4)$	$X(5)$	$X(6)$	$X(7)$	$X(8)$	$X(9)$
name	ID	AGE	SEX	Q1	Q2	Q3	Q4	Q5	Q6

Thus, question 6 (Q6) is variable number 9, and SEX is variable number 3. Any variable may be referred to by its sequence number in subscript form or by its name. It may seem laborious to name a hundred
variables, but if the variables are not named, it requires that one determine the sequence number for each variable. A researcher is unlikely to
mistake AGE for SEX when using names, yet it is very easy to miscount
and to specify variable 105 instead of 106. Such errors are often difficult to detect. It usually pays to name variables.
    Note in the preceding example that everything that is input is con-
sidered to be a variable. Thus, ID is the first variable. One would not
normally take the standard deviation of the ID (or take any statistics,
for that matter). In BMDP, variables can be excluded from analysis
with the USE sentence. To exclude ID in the example, we would use
one of the following:

```
USE = 2,3,4,5,6,7,8,9.
```

```
USE = 2 TO 9.
```

```
USE = AGE,SEX,Q1,Q2,Q3,Q4,Q5,Q6.
```

Once a variable has been excluded in this way, it is unavailable for
use except as a label. Thus, one could label the data for each case by
specifying this:

```
LABEL = ID.
```

or this:

```
LABEL = 1.
```

Of course, if a piece of information has been omitted by writing a
format specification that ignores it, then that information is totally
unavailable. Students encounter difficulty when the number or the

sequence of variables in the FORMAT, VARIABLE, NAME, USE, and LABEL sentences do not add up logically. Look at the following correct examples that use the data from the preceding example:

```
VARIABLE = 9.
FORMAT = '(1F3.0,3X,1F2.0,3X,1A1,3X,6(1F1.0,4X))'.

NAMES ARE ID,AGE,SEX,Q1,Q2,Q3,Q4,Q5,Q6.
USE = 2,3,4,5,6,7,8,9.
LABEL = 1.
```

or (ignoring ID):

```
VARIABLES = 8.
FORMAT = '(6X,1F2.0,3X,1A1,3X,6(1F1.0,4X))'.

NAMES ARE AGE,SEX,Q1,Q2,Q3,Q4,Q5,Q6.
USE = 1 TO 8.
```

If we included a LABEL = 1 instruction in the second example, it would cause BMDP to use the age of each case as a labeling variable. As noted above, the more common mistake is to have disagreement between these sentences, such as nine variables and eight names, which will always cause some form of error.

Note in the last example that the first two sentences and the last two are not grouped together. The reason is that they are in separate paragraphs. Sentences may appear in any order within a paragraph, but their appearance in the wrong paragraph will always result in an error. The VARIABLE and FORMAT sentences appear in the /INPUT paragraph; the NAMES and USE sentences are in the /VARIABLE paragraph. Thus, we can write:

```
/INPUT VARIABLES = 8.
FORMAT = '(6X,1F2.0,3X,1A1,3X,6(1F1.0,4X))'.
/VARIABLE NAMES ARE AGE,SEX,Q1,Q2,Q3,Q4,Q5,Q6.
USE = 1 TO 8.
```

Each paragraph begins with a slash and has a particular name. Within each paragraph are one or more sentences that always end with a period.

BMDP program control language (PCL) is always begun and ended with special paragraphs. The end is indicated by /END, reasonably enough, but there are no sentences (and thus no periods) in this paragraph. The first paragraph is always the /PROBLEM paragraph, and it always has a single sentence—the TITLE sentence. Titles are normally placed within apostrophes and therefore may contain any number of blanks and special characters, as well as the usual alphanumerics. BMDP places a length limit of 160 characters on titles, and blanks count as characters. A title need not be contained within apostrophes if it begins with a letter and contains no blanks or other special characters. As a rule, put all of your titles within apostrophes so that nothing can go wrong. For the preceding example, we would have the following:

```
/PROBLEM TITLE = 'FRESHMAN CLASS (79) SURVEY'.
/INPUT VARIABLES = 9.
FORMAT = '(1F3.0,3X,1F2.0,3X,1A1,3X,6(1F1.0,4X))'.
/VARIABLE NAMES ARE ID,AGE,SEX,Q1,Q2,Q3,Q4,Q5,Q6.
USE = 2 TO 9.
/END
```

Note that the format specification itself is also within apostrophes. There are other specific requirements for BMDP formats; see Chapter 6 and Section 9.2.

The /FINISH paragraph is the source of much difficulty for many students. It is easy if you remember the rule for its use. The /FINISH paragraph is always paired with the CASES sentence of the /INPUT paragraph. If there are 192 subjects in a given experiment, then naturally CASES = 192. However, sometimes one does not know the number of cases or else one suspects that there may be an unknown number of extra cards in the data deck. In this situation, neither the CASES sentence nor the /FINISH paragraph is used. If it does not find a CASES sentence, BMDP will not expect a /FINISH paragraph and will instead read data records until an end-of-file instruction is encountered. (Recall that in an IBM installation the end of the file is indicated by a /* JCL instruction.) If BMDP were being used to list an unknown data set, a laborious counting of cases could be avoided with this option. However, why go to the trouble of a CASES sentence and a /FINISH paragraph if they are not required? Without the /FINISH paragraph, BMDP is never quite sure whether or not it has completed your problem. It is perfectly possible to make a logical error in the PCL instructions such that the program will search for information until either the time allotment expires or the computer operator intervenes. In either case, one has wasted a fair amount of expensive computer time because the computer did not know when to quit. The general principle is to always specify CASES and /FINISH except when using the special "clean-up" facilities of BMDP. A nice feature of the clean-up programs is that they will count your cases for you!

Table 9.1 provides a set of guidelines for the CASES and /FINISH situation. One final caution is that the /FINISH line is *always* the last PCL instruction; i.e., it comes after the data and before the SCL instructions that end your job.

TABLE 9.1. Guidelines for CASES *and* /FINISH *with BMDP*

1. If CASES is specified, /FINISH must be included.
2. The position of the /FINISH paragraph is at the end of the data.
3. There are no sentences in the /FINISH paragraph, so that paragraph *does not* end with a period.
4. Always specify CASES and /FINISH except when using the special clean-up programs (such as P4D).
5. If CASES is specified incorrectly, an abnormal termination of the job will occur, and the amount of usable output will depend on the individual program.
6. When a BMDP tape or disk file is read, the CASES and /FINISH specifications are not used.

The following guidelines apply only to BMDP-81:
7. When /FINISH is not used, a system card (marked /* on most IBM systems) or an /END (note that there is no space between the slash and END) is substituted.
8. The ?FINISH instruction is the command to end an interactive (terminal) session.
9. It is still recommended that you specify CASES for its error-detecting features.

9.2   Input to BMDP

In Chapter 6, we discussed data input procedures in the abstract, without reference to a specific package. Each package is different in its data input procedures, although they should seem familiar even to the beginner.

BMDP has several different modes of input for both terminals and cards. We will begin with the basic procedures learned in Chapter 6 because, once mastered, they will allow you to use any version of BMDP with terminals or with cards. As discussed in the introduction to this chapter, the newer procedures are easier to learn but are still susceptible to simple errors. They do not, however, provide the beginner with insight into the structural arrangement of variables, and, most importantly, they provide little in the way of checks and validations of the input data.

In the applied section (Section 9.3), examples of all BMDP input procedures are provided. Experience suggests that the advanced user should review what is germane, while the beginner should learn the standard input first.

9.2.1   Basic Input Procedure

The basic input procedure for BMDP is to type or punch a FORMAT sentence in the following form:

```
FORMAT IS '(insert any correct specification here)'.
```

We might have any one of the following in an analysis:

```
FORMAT IS '(3X,2F4.2)'.
```

```
FORMAT IS '(5F9.0,1A1/3X,10F2.0)'.
```

```
FORMAT IS '(13F1.0)'.
```

```
FORMAT IS '(4X,2A2,7(4X,3F2.1))'.
```

The following guidelines review the critical aspects of basic BMDP input:

1. The word FORMAT need not begin in column 1. Indentation of all sentences, including FORMAT sentences, makes for better readability.
2. IS, ARE, and the equals sign (=) may be used interchangeably. Thus, we could have:

   ```
   FORMAT = '(3F10.2)'.
   ```

3. The blanks that precede and follow IS and ARE are necessary to distinguish those words from something named FORMATIS or FORMATARE. Blanks are not required for the equals sign.
4. The word FORMAT must be spelled out completely. Other BMDP control options may be abbreviated, depending on the version in use.
5. The word FORMATS is not acceptable.
6. Apostrophes before and after the format specification are mandatory.
7. The period ending the sentence is also mandatory.
8. In most circumstances, blanks are ignored.
9. F, A, and X are the standard format specifications. (See Section 9.4.1 for others.)
10. Special characters other than parentheses, the period, the comma, and the slash are not allowed.
11. If more than one record is required to complete a specification, continue onto the second record as if it were one long record. Only one period is required. The additional format instructions need not begin in a specific column.
12. One FORMAT per analysis is allowed.

13. The instruction might or might not be located in the usual position in the PCL.

With the addition of FORMAT, the apostrophe, and a period to make a sentence, the BMDP basic input is exactly as we learned it in Chapter 6. The following is a list of frequent errors:

missing apostrophe preceding or following a parenthesis;
standard (double) quotation marks substituted for an apostrophe;
final period omitted;
extra or missing delimiter, such as '(3x,,2F4.0)' or '(3X2F4.0)';
impossible, illegal, or illogical specifications, as in '(3X)' and '(3X,5F1)';
a FORMAT written as it would be in the FORTRAN programming language;
punched decimals either not accounted for by the FORMAT or unintended in the first place (a punched decimal always takes precedence).

Three additional common errors associated with input specifications but not, strictly speaking, errors in syntax are as follows:

blank, incomplete, or otherwise imperfect records or SCL or PCL instructions mixed in with the data;
lack of agreement between the data, the FORMAT sentence, the variable count, and the assignment of names to variables, all of which must agree with each other;
specifying the wrong data or the wrong columns (this is not an insult to your intelligence—it happens all the time!).

The entire FORMAT process is omitted when BMDP files are input. Those of you who are using terminals instead of cards are still required to write FORMAT sentences and to arrange each data line as if it were on a card. This is somewhat awkward and defeats many of the advantages of using a terminal. The more recent versions (BMDP-81) allow other forms of input.

9.2.2   The FREE Formats

The new formats are collectively known as FREE formats, and they enable the user to input data without predetermining an exact location. A single line or card may contain many cases or only a part of one case. All such formats use commas or blanks to separate each piece of data.

BMDP simply uses the number of variables input, the number of variables specified, and the number of variable names to determine the organization of the data, thus freeing the user from a frequently boring task. The simplicity of the FREE format procedures is a trap for the beginner, since there are no safeguards. With standard formats, errors are detected at two levels. FORTRAN compilers are able to recognize a number of errors before analysis. In addition, syntactically correct but locationally incorrect formats are often revealed when they produce impossible or illogical results. In a FREE format, neither of these checks exist. But if organizing data and submitting them to a computer are old hat, the FREE formats can and will save time and effort.

The first of the FREE formats is as stated above, with the following rules:

1. Each case begins on a new record (line or card).
2. With multirecord cases, each group must begin on a new line or card.
3. All cases must have the same number of records.
4. All data will be read; i.e., data cannot be ignored.
5. An incomplete record must contain only blanks after the end of the data.
6. Commas or blanks must separate each variable; they may be mixed, but it is better not to do so.
7. Missing values may not be indicated by blanks. The standard missing indicator is the asterisk (*). One asterisk is sufficient to denote a multicharacter variable. Successive commas also serve this function.
8. Decimal values must be punched (or typed) as such.

The following are examples of data records. They illustrate the use of blanks, commas, asterisks, successive commas, and decimals. The data are not from any particular study. This is a straightforward example:

```
010,17,16,74.91,13964,2
174,18,17,66.01,09.78,1
921,15,16,50.66,07.00,2
```

Here is one with missing data:

```
010,*,16,74.91,13964,2
174,18,17,66.01,09.78,1
921,15,16,50.66,*****,2
```

The following example uses blanks instead of commas:

```
010 * 16 74.91 13964 2
174 18 17 66.01 09.78 1
921 15 16 50.66 ***** 2
```

This one shows data requiring multiple records:

```
(record one)
010,17,16,74.91,13964,2,1,1,2,3,4,5,7,84,3,7,4,8,8,7,6,7,3,7,3,7,3,3,4,1,2,4,8,,

(record two)
1,010,2,7,4,3,4,8

(record three)
174,18,17,6.01,9.78,1,3,8,7,6,7,8,9,8,7,8,6,5,5,4,9,8,7,7,8,9,7,6,6,6,9,7,8,7,7,

(record four)
174,2,9,8,8,7,9
```

As you may easily observe from these examples, the FREE format is a very powerful option that requires less time and less planning than most. A dropped comma is not a problem if a blank is used in its place. But when a separator is dropped (as in column 40 in record one), BMDP will not notice. Thus, case 010 would have 35 questions instead of 36. One would be unlikely to notice that the first line contained only 29 questions instead of 30. Terminals also cannot be set to skip at the proper moment to give the user a partial indication of error. So the speed and the ease of data entry are offset by the greater difficulties in error detection.

The problems associated with the FREE format also apply to the STREAM and SLASH formats. A discussion of the error potential of the latter two types of format would be redundant, but it is suggested that you think about the consequences before using them.

9.2.3 The STREAM Format

The rules for the STREAM format can be summarized as follows:

1. Cases may begin anywhere and usually follow each other consecutively.
2. Special indicators are not used for the end of a case, file, etc.
3. Commas or blanks are used to separate each variable.
4. Blanks must not be used for missing values. The asterisk is the standard indicator of a missing value.
5. Each line or card must be completed, and there is no way to ignore data.
6. End-of-data and end-of-file conditions are detected.
7. Divisions between cases are determined only by the variable count.

Consider a short survey in which the data to be input consist of an ID and the answers to 12 questions. With three-digit ID's and coded answers of yes (1), no (2), undecided (3), and unanswered (4), we might have these records:

```
001,1,2,2,1,2,2,1,1,1,2,2,1,002,1,1,1,1,1,2,2,2,1,2,1,1,003,1,1,2,2,2,1,1,1,1,2,
1,2,004,2,1,1,1,*,1,1,2,4,*,1,005,1,2,1,2,2,2,2,1,1,2,1,1,006,1,1,1,1,2,2,1,2,2,
1,2,2,007,1,1,1,1,2,2,1,1,2,2,1,1,008,1,2,1,2,1,2,1,2,2,2,1,1
```

Note that the individual cases start in a variety of places on the records. Proofreading is very difficult under these circumstances, and it is impossible to check for errors other than "bad" characters. With this form of input, checking and cleaning of data are confined to manual proofing and to the use of the programs devoted to this area, e.g., P4D.

Dropping a datum here has very serious effects. If the last question in case 004 above is omitted, then the next ID (005) becomes the answer to that question, and 1, the first answer in that case, becomes the ID. This will be discovered only if the investigator remembers to set minimums and maximums for each variable and thus detects the 005 as an unintelligible answer. Be cautious.

9.2.4   The SLASH Format

The third and final format covered in this section, the SLASH format, uses the slash (/) to separate cases, allowing multiple cases per record and multiple records per case. The rules are as follows:

1. Multiple cases may appear in a line (or card), or a case may extend to several lines (or cards).
2. The slash (/) denotes the end of a record.
3. Variables are separated with commas or blanks.
4. There is no way to ignore data.
5. The data must be continuous, i.e., with no strings of blanks.
6. Only "bad" characters are checked for.

In essence, this option is identical to the STREAM format with the addition of a specific end-of-case operator. SLASH is recommended over STREAM because the data from separate cases will not be confused. It is still difficult to check within each case.

9.2.5   Summary

As we have seen, BMDP input consists of a formatted, card-oriented system and a nonformatted, terminal-oriented system. Formatted procedures are slow, somewhat prone to logical errors, and awkward to use on a terminal. They do have two important virtues, however; these are the learning advantages and the proofreading capabilities. It is very useful to the beginning user to learn the structure of formatted systems. Use of the structured systems tends to prevent the "garbage-in, garbage-out" syndrome, and it forces the user to consider the data in an ordered and logical manner. When the user advances to the stage of writing his or her own programs, the structured systems will transfer rather directly. Also, a structured format is easier to read because spaces are allowed.

Many terminal systems add other complicating factors (files, disk space, log on and off, etc.) that add to the burden on the beginner. I do not mean to suggest that the final word is in regarding the relative merits of the two systems. FREE format input procedures are a significant advantage on a terminal, but they are not the place to begin.

9.3   BMDP Applied

At this point, you should be somewhat familiar with the BMDP package. If you have not yet read the introductory material, do so. Remember that the specifics of a program are detailed in the corresponding manual and not in this text. The sections on transformations and files need not be read until you come to Exercises 3 and 4. What this section will do is provide exercises that utilize the options of several representative BMDP programs. The objective is to give you some practice and some examples for future reference.

9.3.1   Practice with Program P4D

P4D is the place to begin learning BMDP. This program requires fewer PCL statements and is designed to clean up data before analysis. It is a record-oriented program, which means that files cannot be created or retrieved. The difference between BMDP and BMDP Versions 77, 79, and 81 is the allowable abbreviations and specifications in the /PRINT paragraph. These names and instructions are not interchangeable, and the paragraphs and sentences should be placed in the order given in the manual.

━━━━━ **EXERCISE 1** ━━

Given either terminal or card input, write the program control language (PCL) for the transportation survey data in Table 9.2. Check all columns for improper data. Produce a copy of the raw data, and substitute the vertical bar (logical OR) for all numbers. Five cases are given in Table 9.2; read the descriptions of P4D, and write the PCL needed to analyze those cases.

TABLE 9.2. Transportation-Survey Practice Data

	Data Description		
Variable	Number of digits	Number of decimal places	Function
1	3	0	identification
2	1	0	record number
3	2	0	age
4	1	0	gender
5	1	0	SES
6	2	0	voting district
7	1	0	number of family automobiles
8	2	0	miles to work for husband
9	2	0	miles to work for wife
10	2	0	public transportation used by husband
11	2	0	public transportation used by wife
12	5	0	total miles traveled by husband's car
13	5	0	total miles traveled by wife's car
14	5	0	total miles traveled by car 3
15	5	0	total miles traveled by car 4
16	5	0	total miles traveled by car 5
17-30	1	0	survey questions 1–14

Raw Data

```
001 1    29 M 4 31    2 11 00 05 00 19000 09700 00000 00000 00000    8 6 3 3 4 5
001 2    4 3 7 2 7 8 3 3
002 1    41 M 2 22    2 27 09 00 10 25000 10000 00000 00000 00000    7 9 6 7 7 8
002 2    9 6 9 8 6 7 8 9
003 1    19 M 5 03    1 12 00 05 40 09000 00000 00000 00000 00000    2 4 1 3 5 6
003 2    2 3 2 4 1 1 7 1
004 1    24 M 4 17    2 07 09 25 00 06000 10000 00000 00000 00000    1 2 5 3 4 4
004 2    2 2 1 1 5 4 6 1
005 1    35 M 3 02    1 00 15 99 00 00000 14000 00000 00000 00000    2 2 3 2 4 4
005 2    1 1 2 3 6 2 1 2
```

Answer:

```
/PROBLEM TITLE 'PUBLIC VS PRIVATE TRANSPORTATION SURVEY'.
/INPUT VARIABLE =   160.
FORMAT  = '(80A1/80A1)'.
/PRINT NUMERIC  = '|'.
ILLEGAL = '*'.
/END
```

The answer to Exercise 1 does not differ much in concept from that to Exercise 2, which follows. Accordingly, the discussion of these two has been merged and placed with Exercise 2.

========= **EXERCISE 2** =========

Consider an insomnia-reduction study with two factors. Factor **A** is the treatment factor, which has three levels: experimental treatment, attention/placebo, and no treatment. Factor B is testing before and after treatment. The dependent measure is the average number of hours slept per night. If you are unfamiliar with multiple-factor studies, ignore Factor B and consider this to be a one-factor design that uses only the post-treatment test information. (*Note:* Because we are trying to keep this simple, we are not availing ourselves of the most desirable design for this type of experiment. Huck and McLean[1] suggest not using before and after as a factor. See their article for recommendations.) Further, assume that the variables are on the terminal lines or cards described in Table 9.3. Read the description of P4D, and write the PCL required to analyze the cases.

Answer:

```
/PROBLEM   TITLE =  'INSOMNIA TREATMENT'.
/INPUT  VARIABLE =   80.
FORMAT = '(80A1)'.
/PRINT NUMERIC = '|'.
ILLEGAL = '*'.
/END
```

The first thing that one notices about these two exercises is their striking resemblance to each other; only the problem titles differ. In essence, all P4D programs are similar no matter what the data or the ultimate goal of analysis. Even the /INPUT paragraph could have been omitted because each column is a variable (and so we have 80 variables), and so the format is assumed to be 80A1. It serves no purpose to assign names to the variables, since few are one-column variables. If we included a specification such as /VARIABLE NAMES = ID,TREATGRP,SEX,AGE, etc., then column 1 would be labeled ID, column 2 would be TREATGRP, column 3 would be SEX, and so on. This would be a total mislabeling of variables; hence, it is best not to name variables in P4D.

In multiple-record situations, it is a good idea to specify all columns as variables. This provides a guarantee that blank columns will indeed be filled with blanks. P4D does have a limit of 200 variables, or $2\frac{1}{2}$ cards, if all columns are used. In cases where only 50 columns of every record are occupied, there could be as many as four cards per case. For large numbers of records per case, specify blocks of columns and make multiple passes through the data.

It is suggested that you enter the number of variables and the format used for practice and learning purposes if you are a beginner. NUMERIC = '|' was chosen in

---

[1] Huck, Schuyler W., and Robert A. McLean. "Using a Repeated Measures ANOVA to Analyze the Data from a Pretest-Posttest Design: A Potentially Confusing Task," *Psychological Bulletin,* 1975, 82 (4), pp. 511–518.

TABLE 9.3. Insomnia-Reduction Practice Data

| | Data Description | | |
| | Number of | Number of | |
Variable	digits	decimal places	Function
1	2	0	identification
2	1	0	treatment group
3	1	0	gender
4	2	0	age
5	2	0	insomnia type
6	3	1	pre-treatment hours slept
7	3	1	post-treatment hours slept

Raw Data

```
01 1 1 22 07 050 075
02 1 2 17 01 043 081
03 1 2 18 03 038 045
04 2 1 19 03 044 043
05 2 2 19 07 051 055
06 2 2 18 04 030 038
07 3 1 21 01 041 071
08 3 1 20 08 055 065
09 3 2 20 10 025 045
```

both answers because when all numbers are replaced with the vertical bar in the printed output the result will be neat columns that can be easily scanned. In the above cases, the M's and F's would remain. The end result is a matrix of cases by record columns. Incorrect substitution of letters for numbers and similar errors will stand out. This rapid and facile checking of the data preparation is the central function of P4D. (These features apply to both terminal and card data entry.)

9.3.2   Practice with Program P1D

Though P1D can be used with a file that has been saved, it is very rarely done. It is common to use P1D to create a file for the first time. Other programs can be used to create files, but the task is easier when the data are being checked and cleaned. Accordingly, build a BMDP file in Exercise 3 and retain it for further use. In Exercise 4, transform the data (create a new variable). Do not infer that political scientists and sociologists always use files and never transformations and that psychologists do the opposite. The topics are presented separately for clarity, and practice with both is highly recommended. In Exercises 3 and 4, name your variables, set maximum and minimum values for each, and print the data.

===== EXERCISE 3 =====

In addition to fulfilling the above requests, define the meaning of a blank in the data by using the BLANK sentence. Do not attempt transformations at this time. (Also skip the ADD, BEFORE, USE, CONTENT, and NEW procedures.) Use the data in Exercise 1.

Answer:

```
/PROBLEM TITLE = 'PUBLIC VS. PRIVATE TRANSPORTATION SURVEY'.
/INPUT VARIABLES = 29.
FORMAT = '(1F3.0,7X,1F2.0,1X,1A3,1X,1F1.0,1F2.0,3X,1F1.0,1X,4(1F2.0,1X)5(1F5.0,
1X)3X,6(1F1.0,1X)/10X,8(1F1.0,1X))'.
CASES = 5.
/VARIABLE NAMES = ID,AGE,SEX,SES,VOTDIST,NUMAUTO,MILESHUS,MILESWIF,PUBTRANH,
PUBTRANW,MIAUTOH,MIAUTOW,MIAUTO3,MIAUTO4,MIAUTO5,Q1,Q2,Q3,Q4,Q5,Q6,Q7,Q8,Q9,
Q10,Q11,Q12,Q13,Q14.
MAXIMUM = 5,40,2,9,50,5,200,200,99,99,40000,40000,20000,20000,20000,9,9,9,9,9,9,
9,7,7,7,7,3,3,3.
MINIMUM = 1,10,1,1,01,0,000,000,00,00,00000,00000,00000,00000,00000,1,1,1,1,1,1,
1,1,1,1,1,1,1,1.
BLANK = MISSING.
MISSING = (16)0,0,0,0,0,0,0,0,0,0,0,0,0,0.
/SAVE CODE = DMKTRANS.
LABEL = 'PUB VS PRI TRANS SURVEY FILE ONE'.
UNIT = 7.
/PRINT MISSING.
MAXIMUM.
MINIMUM.
DATA.
/GROUP CODE(3) = 1,2.
NAMES = HUSBAND,WIFE.
CUTPOINTS(2) = 35.
NAMES = '35 AND UNDER','OVER 35'.
/END
```

A number of issues are illustrated in this answer. For clarity, we will take them up in the order created by the answer itself. (Remember that the odd-numbered exercises attempt to match the typical needs of the political scientist and the sociologist, while the even-numbered ones are for the psychologists and the more laboratory-oriented social scientists. Therefore, we go to the trouble of creating a BMDP system file in the odd-numbered exercises but not in the even-numbered exercises.)

The first paragraph holds no surprises. However, the VARIABLES sentence in the /INPUT paragraph indicates 29 variables instead of the 30 in the data. The record sequence number is the missing variable. If you recall, that number is primarily for the researcher's convenience and serves no purpose in analysis. In short, this variable can be dropped at this point because it will not enter into any statistical analysis. (See the discussion of the OSIRIS system in Appendix C for an interesting exception.)

The FORMAT sentence is unremarkable and should be thoroughly familiar at this point. The CASES sentence is also unremarkable. You may be wondering why we need to specify cases when five is such a trivial number. It is very important to get into the habit of always specifying cases because this practice can prevent erroneous results and abnormal operation of BMDP itself. If your data deck contains too few or too many cards or too many terminal lines, BMDP will inform you of this, but only if CASES is specified. Since CASES always requires a /FINISH paragraph, the CASES instruction will also prevent BMDP from wasting your valuable account time in a search for more information. This happens when CASES and /FINISH are not specified and when other sentences do not form a logical end to the PCL. The process will halt when the automatic time default expires or when the computer operator notices that your program is looping (stuck) and summarily cuts you off!

The NAMES sentence contains nothing unusual. Note that there is no variable for the card number. The MAXIMUM sentence lists the maximum for each variable in sequence. This is an inclusive maximum; the value is stated. For example, 5 for NUMAUTO includes the number 5 as a possibility. If there were no reason to state the MAXIMUM for the questions (Q1 to Q14), then the list could end with the 20,000 limit set for variable MIAUTO5. A perfect correspondence between the NAMES list and the MAXIMUM list is required to use this form. Thus, if you state a maximum for your last variable, you must do it for all variables.

The way to avoid stating a maximum for all variables is to use the BMDP tab feature. The BMDP manual suggests an analogy between this feature and typewriter tabulation. On a typewriter, tabs could be set for every ten spaces, for example, to correspond to ten variables. However, in the BMDP tab feature, the number of columns occupied by the variable and the sequence are not relevant. In this answer, NUMAUTO is variable 6, and we could state the maximums for it as follows: MAXIMUM = (6)5. Setting maximums for AGE, VOTDIST, NUMAUTO, and Q14, we have this:

MAXIMUM = (2)40,(5)50,5,(29)3.

or this:

MAXIMUM = (2)40,(5)50,(6)5,(29)3.

Thus, in defining the maximums for these variables, BMDP tabs to the variable, makes the assignment, and moves on to the next variable. Note that when variables follow in sequence, the sequence or the variable number need not be stated.

Variable 6 is next in sequence after variable 5. Unlike a typewriter, BMDP can tab in both directions or to the same variables multiple times. The last assignment is the one that BMDP will use. If you tab out of order, it will use computer time need-lessly, and, more importantly, this may be a source of error. Generally speaking, this practice is not advisable.

The tab feature can be used with any list—not just with a list of maximums. If we were to write a name sequence naming only AGE, VOTDIST, NUMAUTO, and Q13, the following would be a standard solution:

```
NAMES = (2)AGE,(5)VOTDIST,NUMAUTO,(28)Q13.
```

A blank inserted before or after these names would violate the conventions for variable names.

These procedures also apply to the MINIMUM sentence, as one would expect. There is one other feature of BMDP that is quite useful for maximums and is very handy for minimums. This is the repeated, or multiple, assignment of numbers, which can be combined with tabbing. Examining the MINIMUM sentence in this answer, you can see that there are several series of variables, all having the same minimum. The sentence could be shortened considerably, in the following two ways:

```
MINIMUM = 1,10,1,1,01,0,000,000,00,00,5*00000,14*1.
```

```
MINIMUM = 1,10,2*1,01,0,2*000,2*00,5*00000,14*1.
```

Of course, it is a bit ridiculous to use a repeated assignment for only two repeats, but it illustrates the point.

Skipping the BLANK sentence for the moment, we come next to the MISSING sentence. The MISSING sentence allows the user to specify a code for missing values in much the same way as MAXIMUM and MINIMUM sentences do. Thus, in our example, we have tabbed to variable 16, assigned it the code 0, and then (in se-quence) assigned 0 to variables 17–29. (These are Q1 through Q14.) Assignments could have been made for the first 15 variables if desired. However, in the case of variable 11 (miles driven in car number 1), a code of 0 for a missing value is inap-propriate in that if a person does not own a car, the data are not missing. Rather, the figure of 00000 miles should be used. For this reason, only the questions (Q1 through Q14) are assigned zero as the missing value. Using all of this, we can reduce the MISSING sentence to this:

```
MISSING = (16)0,13*0.
```

There is no means of voiding a missing value code within a given run.

Taking up the BLANK sentence, we can now see that it has an interrelationship with the MISSING sentence. In this example, blanks are assumed to indicate that those variables having logical and possible zeros are missing. Both of the sentences interact with the MAXIMUM and MINIMUM sentences in that what is assumed missing should not also be the maximum or the minimum value. It is a wise precaution, if the program allows it, to have all information about maximum, minimum, and missing variables printed as part of the output. In this way, you will know exactly what was eliminated from any given analysis. Be cautious.

There is, in this author's experience, a troublesome difference between older and newer BMDP versions with regard to the BLANK sentence. In BMDP, there is no option, and BLANK always equals zero. But in the newer packages, BLANK is also allowed to equal MISSING. In all packages, there is no direct procedure for nonnumeric codes for missing values. Alphabetic and special characters must be converted with the TRANSFORM option, or else they will be excluded from all analyses.

The next paragraph is /SAVE, and it retains the data as part of the internal computer memory, thereby facilitating future analysis. There is nothing magical about using Program P6D to do this; any BMDP program except P4D could be used. The essence of saving the data in a BMDP system file is the proper assignment of sufficient identification. The file must be distinguished from all others that might be present. It is also important to note that a BMDP system file is a public file, so you will be sharing it with many others. This does not mean that anybody can gain access to your data, however. To illustrate this point, consider the CODE sentence in this answer. It names the file DMKTRANS. Thus defined, the file is inviolate as long as no one knows its name. That is, it appears to each user of the BMDP system file that his or her file is the only one in existence. Only those users who know the name of a file can gain access to it.

File codes are limited to eight alphanumeric characters. If we wish to create several files coded DMKTRANS, the CONTENT sentence is required. This is a further expansion of the file-naming process. A file may also be given a LABEL (of less than 40 characters) to further identify it. Only CODE is really necessary, though. In the answer, we can see that both CODE and LABEL have been used.

The next sentence is the UNIT sentence, which is mandatory and which may not specify 1, 2, 5, or 6. This is an internal programming issue that need not concern the user. Exactly which unit will be used is determined by the computer center personnel when BMDP is installed. The units chosen may or may not provide for options. Remember that the system control language (SCL) will establish where these files are.

The final option in the /SAVE paragraph is the NEW sentence. Caution must be used with this because if you specify NEW for a file with information already in it, that file will be erased. This is particularly important when you are using public files. You will usually be adding your data to the file, and the NEW sentence will not be needed. To use a file as input, you will use the same procedure except that a FORMAT statement must not be used (see Exercise 5) and the CODE, LABEL, CONTENT, and UNIT sentences become part of the /INPUT paragraph. Thus, we use /SAVE to make a file and /INPUT to retrieve one.

Next is the /PRINT paragraph, which enables you to print missing, maximum, and minimum values as well as the data themselves. This is an option because once the data are properly purged of "bad" information, it wastes time and paper to print them. If you specify all options, it provides a list of cases containing missing, maximum, and minimum occurrences. The data listing can then tell you what the offending numbers actually are. *Note:* P1D is not designed to help you locate bad or inappropriate characters.

The last paragraph is /GROUP, and it is used for subdividing the data into groupings. In the answer, we divided the data into one subset containing interviews with husbands and one subset of interviews with wives. The NAMES sentence does just that, but it must correspond in sequence to the CODE sentence. If the code is reversed (i.e., 2,1) or if the names are reversed (WIFE,HUSBAND), the output will be mislabeled. The CUTPOINTS sentence serves the same purpose for continuous variables as CODE does for discrete variables. A CODE sentence could have been written for AGE except that each and every year would then become a subset, as in the following:

```
CODE(2) = 1 TO 99.
```

Values outside of the specified codes or cutpoints are excluded from analysis, but, unlike MAXIMUM and MINIMUM procedures, with these BMDP will not note what has and has not been excluded. In general, the /GROUP paragraph should not be used for case inclusion or exclusion but only for subsetting.

The final paragraph is, of course, /END.

═══════════ EXERCISE 4 ═══════════

After Exercise 3, many parts of Exercise 4 seem redundant. However, the /TRANSFORM option is important to understand, and the discussion should be read by survey-oriented people as well as by experienced users who want review and practice.

The experimental design of this exercise is a 3×2 factorial experiment concerned with the effects of a therapeutic practice compared to controls. Using the data from Exercise 2, write the PCL for P1D and add a new variable—the difference between hours slept before the test and the hours slept after the test. Refer to this new variable as CHANGE.

Answer:

```
/PROBLEM TITLE = 'INSOMNIA REDUCTION'.
/INPUT VARIABLES = 7.
FORMAT = '(1A2,1X,2(F1.0,1X)2(F2.0,1X)2(F3.1,1X))'.
CASES = 30.
/VARIABLE NAMES = ID,TREATGRP,SEX,AGE,INSOMTYP,PREHOURS,PSTHOURS,CHANGE.
```

```
MAXIMUM = 15,3,2,99,11,10,24.
MINIMUM = 01,1,1,1,15,01,00,00.
BLANK = MISSING.
MISSING = (3)9,00,99,999,999.
ADD = 1.
/TRANSFORM CHANGE = PSTHOURS - PREHOURS.
/PRINT MISSING.
MAXIMUM.
MINIMUM.
DATA.
/GROUP CODE(2) = 1,2,3.
NAMES = TREATMNT,APLACEBO,CONTROL.
/END
```

Here, the /PROBLEM and /INPUT paragraphs are not surprising. However, note that labels read in with A-format specifications count in the variable total. Thus, everything is a variable to BMDP—not just the independent and dependent variables. Data that are ignored by the FORMAT statement are not added to the variable total. Remember that in FREE formats it is not possible to ignore anything that is on the line or card.

The /VARIABLE paragraph is as expected except for the last name and a new sentence (ADD), which function together with the /TRANSFORM paragraph to create new variables. If 45 new variables are created, then 45 extra names and the instruction ADD = 45 are required. If new variables are not assigned names, then they must be referred to by subscript. For the present example, CHANGE is also $X(8)$. This is not a recommended procedure, for it requires that you remember or keep a list of transformations. If you take the trouble to assign names, you are unlikely to use the wrong variable or to confuse them. The subscripts assigned begin after the last regular variable and follow in the sequence in which they are created by the /TRANSFORM paragraph.

The transformation is a very simple one, but it illustrates several important points about this paragraph. The arithmetic operation of subtraction is used to create the new variable CHANGE. These operations are stated in standard FORTRAN language; i.e., addition is indicated by the plus sign (+), subtraction is indicated by the minus sign (−), multiplication by the asterisk (*), and division by the slash (/). In the present example, CHANGE is defined as PSTHOURS (hours slept after the test) minus PREHOURS (hours slept before the test). The ratio of the two could be obtained by division. BMDP-77 restricts the user to one operation per line. For example, to state CHANGE in minutes, we would write this:

```
/TRANSFORM CHANGE = PSTHOURS - PREHOURS.
CHANGE = CHANGE * 60.0.
```

or this:

```
/TRANSFORM PSTHOURS = PSTHOURS*60..
PREHOURS = PREHOURS * 60.0.
CHANGE = PSTHOURS - PREHOURS.
```

In effect, we are not allowed to multiply and subtract in the same expression. Also, note that the decimal in the multiplication of minutes will not serve as the period at the end of a sentence. Insertion of a zero after the decimal is recommended, but not required.

BMDP-77 is able to use most of the operations associated with standard FORTRAN, including trigonometric functions such as sines and cosines. Logical operators are also possible; they are very useful in case selection and deletion. A special USE sentence is added to the /TRANSFORM paragraph to instruct BMDP about what should be done with the results of a logical operation. This is not the same as the USE sentence and /VARIABLE paragraph combination. The following transformation deals only with those subjects who gained more than one hour of sleep.

```
/TRANSFORM CHANGE = PSTHOURS - PREHOURS.
USE = CHANGE GE 1.0.
```

Only when the value of CHANGE is one hour or more is USE true, i.e., positive. Whenever USE is zero (false) or negative, the result is that the value of CHANGE is not used. GE is the logical operator "greater than or equal to." Hence, if PREHOURS were 4.5 and PSTHOURS were 7.0, CHANGE would equal 2.5. This is obviously greater than 1.0, causing USE to be considered true. This transformation would be executed for each subject excluding those who gained less than one hour. Subjects with negative gain (sleeping less after the treatment) would also be excluded.

BMDP-81 contains many revised and improved transformation capabilities. The most obvious change is the elimination of the one-operation-per-line restriction. Thus, the example changing hours to minutes can be written in one instruction:

```
/TRANSFORM CHANGE = (PSTHOURS - PREHOURS) * 60.0.
```

Transformation instructions written as in the earlier BMDP versions will continue to function correctly, but they are rather inefficient. There are many other options available in the /TRANSFORM paragraph, but they are somewhat specialized and should be considered advanced topics. These options are reviewed in the "Pro Tips and Techniques" section.

The various requests in the /PRINT paragraph are self-explanatory. The DATA section is very useful in that it will provide a listing of your data in the original input arrangement; they will not be arranged in groups of ten columns as P4D does. When things go wrong, it is much easier to check the pages of output of P1D than to sift through hundreds or thousands of terminal lines or cards.

The /GROUP paragraph allows data to be organized by treatment. The order of codes is arbitrary, but the name sequence must correspond. For example, the paragraph could read as follows:

```
/GROUP CODE(2) = 1,3,2.
NAMES = TREATMNT,CONTROL,APLACEBO.
```

Note that the variable in question is indicated by its sequence subscript and not by name. CODE and NAME pairs may be repeated as necessary. In general, the practice of excluding data by not referring to a CODE is not recommended because you cannot change your mind once the data have been excluded.

The last paragraph, /END, should be familiar by now.

Exercises 3 and 4 contain a message beyond the details of BMDP. The goal in using a statistical package is to pick your way through the maze of options without creating a logical impossibility and not merely to choose an appropriate analysis technique. As an example of an incorrect choice, we might eliminate a variable in the /TRANSFORM paragraph and then attempt a /GROUP on that variable. This situation may seem unlikely, but it happens all the time.

9.3.3   Practice with Program P4F/P1F

The next exercises are the first that call for an actual analysis with BMDP. Since sociology and political science differ from psychology in typical analyses, Exercises 5 and 6 request different programs. Again, working through both will add to your general experience with BMDP. *Note:* Those readers who are working with BMDP-81 should use program P4F, which replaces P1F, P2F, and P3F of the earlier versions. The instructions are identical provided that P4F users refer only to the appropriate sections of the manual.

───────── **EXERCISE 5** ═══

Recall that the data in Exercise 1 are from a transportation survey. Using BMDP program P4F/P1F, obtain the following series of two-way frequency tables with this data:

    AGE by MILESHUS
    AGE by PUBTRANH
    AGE by total annual mileage for the family
    Q5 by MIAUTOH

MILESHUS is the number of miles the husband must travel to work. PUBTRANH equals the percentage of public transportation used by the husband in commuting to work. The total annual mileage for the family is a variable obtained by summing the mileage of the autos driven by a family. Q5 is a question asking, "Would you consider van-pooling if it were available in your area?" The responses are coded (1)

No, (2) Perhaps, and (3) Yes. MIAUTOH is the number of miles annually that the auto was used by the husband to commute to work. Note that one of these variables must be derived and that the intervals for counting frequency are not built in. If the husband's mileage can vary between 00000 and 50,000, then there are 50,001 potential categories. In this case, use CUTPOINTS.

Assume that the data are in a file named DMKTRANS and are labeled 'PUB VS PRIVATE SURVEY FILE ONE.' Remember that you must specify the unit in which the file is located (this is arbitrary here).

To simplify the problem, ignore the following P1F options:

/TABLE   CONDITION, COUNT, DELTA, MINIMUM
/CATEGORY   RESET, COEFFICIENT
/PRINT   ROWPERCENT, COLPERCENT, TOTPERCENT, DIFFERENCE,
    STANDARDIZED, ADJUSTED, FREEMAN, SMOOTHED

This does not imply that the above options are not useful. Rather, the wealth of options masks the essential points.

Lastly, use the /STATISTICS option to obtain a chi-square test for Q5 by MI-AUTOH, but do not use this option in obtaining the other frequency distributions.

Answer:

```
/PROBLEM TITLE = 'PUB VS PRIVATE TRANS SURVEY FREQUENCIES AND CHI-SQ'.
/INPUT CODE = DMKTRANS.
LABEL = 'PUB VS PRIVATE SURVEY FILE ONE'.
UNIT = 7.
/VARIABLE ADD = 1.
/TRANSFORM TMIA = MIAUTOH + MIAUTOW.
TMIB = MIAUTO3 + MIAUTO4.
TMIB = TMIB + MIAUTO5.
TOTFAMMI = TMIA + TMIB.
/TABLE ROW = AGE,AGE,AGE,Q5.
COLUMN = MILESHUS,PUBTRANH,TOTFAMMI,MIAUTOH.
/CATEGORY CUTPOINTS(2) = 20,30,40,50.
NAMES(2) = UNDER21,UNDER31,UNDER41,OVER51.
CUTPOINTS(7) = 05000,10000,15000,20000,25000,30000,35000.
NAMES(7) = '5000 OR LESS','5001 TO 10000','10001 TO 15000','15001 TO 20000',
'20001 TO 25000','25001 TO 30000','30001 TO 35000','OVER 35000'.
CUTPOINTS(9) = 20,40,60,80.
NAMES(9) = '20 PERCENT OR LESS','21 TO 40','41 TO 60','61 TO 80','81 TO 99
PERCENT'.
CUTPOINTS(30) = 10000,20000,30000,40000,50000,60000,70000,80000,90000.
NAMES(30) = A100RLESS,B10TO20,C20TO30,D30TO40,E40TO50,F50TO60,G60TO70,
H70TO80,I80TO90,OVER90.
CUTPOINTS(11) = 05000,10000,15000,20000,25000,30000,35000.
CODE(20) = 1,2,3.
NAMES(20) = NO,PERHAPS,YES.
/PRINT OBSERVED.
/TABLE ROW = Q5.
/STATISTICS CHISQUARE.
/END
```

The answer as printed here is only a guide, of course, since there are other possible answers. Beginning with the /TABLE paragraph, we can see that the frequency tables were defined by the ROW and COLUMN sentences. Which variable is defined as a ROW or a COLUMN is arbitrary, except that the variables should not be switched within a single run. The tables are formed by the sequence of variables, such as one with one. Note that AGE is not the first variable according to the /VARIABLE paragraph, and if we were to use its subscript instead of its name, we would have

```
ROW = X(2),X(2),X(2),Q5.
```

(Recall that these variable names were assigned when the file was created in Exercise 3.)

All possible pair tables may be generated by specifying CROSS instead of the assumed variable, PAIR. This is a useful but hazardous feature. In our example, if CROSS had been specified, sixteen separate tables would result. Many of these would have no use or meaning. The result is wasted time and money. With 15 ROW and COLUMN variables, CROSS would produce 225 tables!

The /CATEGORY paragraph allows the user to define the number of levels or subcategories of the variables and to label each with a name. As stated in the instructions for this exercise, not defining CUTPOINTS will result in each discrete value becoming a level. Variables are referred to by subscript and not by name in this paragraph. Note that in the NAMES sentences when more than eight characters or special symbols (including blanks) are needed the literal specification is used. That is, the names are set within apostrophes and are separated by commas. The NAMES(30) sentence does not use apostrophes, since no special characters are used and there are eight or fewer characters per name. Each name begins with a letter; this is not for the sake of appearances but because the first character may not be a number when literals are not used. Note that names are not required in such cases as variable 11.

The /PRINT instruction is not really necessary when only the observed (OBS) frequencies are of interest. It has been included here for clarity.

The program would often end at this point, but our problem calls for a chi-square test of Q5 versus MIAUTOH (miles that the husband's auto was driven) but not for all combinations of the variables. By inserting a new /TABLE paragraph before /END, we can request this single analysis without resubmitting an entire job. When working with data in files, considerable savings will be realized by consecutive rather than separate analyses.

The /STATISTICS paragraph is self-explanatory in that it requests just what we need—CHISQUARE. It would appear that this is unnecessary, since CHISQUARE is the assumed option. However, when there are separate analyses, as in this case, the assumption is based on the previous problem, which did not have a /STATISTICS paragraph. The BMDP manuals have very thorough discussions of the other options in the /STATISTICS paragraph. Use them if you understand them; otherwise, they are best ignored. There may be any number of repetitions of the /TABLE to /STATISTICS paragraphs before the /END.

Other sections in P4F extend the basic procedures to apply to situations of empty cells, departures from independence, and multiway frequency tables. The basic form of these procedures is that of two-way tables. The number of options and the statistical expertise required is somewhat greater than that needed for the basic procedures, however. In the older versions of BMDP, these extensions are contained in the separate programs P2F and P3F. However, both the form and the logic of the new and old versions are the same and should not cause problems. (See the "Pro Tips and Techniques" section.)

===== **EXERCISE 6** =====

In many research efforts, it is wise to examine the data for violations of the assumptions underlying the statistics. There are many sophisticated ways to accomplish this, but most of them are well beyond the limits of this text. A somewhat rough-and-ready way to achieve the same basic goal is to use scatter plots and histograms.

Using P6D and the data from the insomnia study in Exercise 2, form scatter plots for pretreatment versus post-treatment, age versus sex, and CHANGE versus age. (Recall that CHANGE was a derived variable from Exercise 4.) Do this for each dependent variable. Also examine age versus sex for all subjects as a whole.

As a final task, use the /STATISTICS option to obtain plots, means, standard deviations, and correlations for CHANGE versus each treatment group and for age versus sex (all subjects). Use caution when interpreting the correlation coefficient with the three-valued distribution of the treatment groups.

Use either a formatted input or a file input depending on what your system allows.

Answer:

```
/PROBLEM TITLE = 'INSOMNIA REDUCTION SCATTER PLOTS'.
/INPUT VARIABLES = 7.
FORMAT = '(1A2,1X,2(F1.0,1X)2(F2.0,1X)2(F3.1,1X))'.
CASES = 30.
/VARIABLE NAMES = ID,TREATGRP,SEX,AGE,INSOMTYP,PREHOURS,PSTHOURS,CHANGE.
MISSING = (3)9,00,99,999,999.
ADD = 1.
GROUPING = TREATGRP.
/TRANSFORM CHANGE = PSTHOURS - PREHOURS.
/INPUT CODE(2) = 1,2,3.
```

```
NAMES = TREATMNT,APLACEBO,CONTROL.
/PLOT XVAR = PREHOURS,AGE,CHANGE.
YVAR = PSTHOURS,SEX,AGE.
CHAR = FREQ.
GROUP = TREATMNT.
GROUP = APLACEBO.
GROUP = CONTROL.
/PLOT XVAR = CHANGE.
YVAR = TREATGRP.
STATISTICS.
/PLOT XVAR = AGE.
YVAR = SEX.
STATISTICS.
/END
```

Note that in this solution we have used the GROUPING sentence to define the independent variables. Once this is established, each /PLOT paragraph creates separate plots for each independent variable or for all groups as one. Thus, the first /PLOT paragraph will produce nine separate graphs by restating the GROUP sentence three times. A single GROUP sentence creates plots having the individual groups superimposed on each other. Also, any statistics will be based on all subjects —not on subjects within groups. When the grouping option is taken, P6D usually uses a separate symbol for each independent variable.

In the second /PLOT paragraph, we obtain CHANGE versus TREATGRP with associated statistics. Since no GROUP sentence is specified, all groups is the assumed value. This means that all subjects are included in the various statistics. Given that TREATGRP is the three-valued independent variable, this plot is a frequency histogram, and it allows us to observe each independent variable's characteristics with respect to CHANGE.

Lastly, the scatter plot of AGE versus SEX is obtained for all subjects. In this plot, each group will be represented by the first letter of its name. So in this case, the graph is comprised of T's, A's, and C's. Remember that the statistics are not based on grouping.

The P6D program can produce a wide variety of scatter plots, and you can tailor them to your specific needs by repeating GROUP sentences and /PLOT paragraphs as needed. Clearly, it is better to do this within a single run than to customize each plot separately. Be cautious about taking the assumed values, for in anything other than very small data sets, a large number of meaningless and wasteful plots will be produced.

9.3.4   Practice with Programs P1R, P1D, and P2V

The final exercises deal with research questions that are more typical of the actual goals of social scientists. In Exercise 7, we consider the question of why people do or do not choose to use public transportation.

In Exercise 8, we study whether or not certain experimental treatments are of aid in improving sleep. It must be emphasized that these questions could be approached in a great many ways and that the choices made here are intended to illuminate BMDP and not research *per se*.

════════ EXERCISE 7 ════════

Assume that the following list of questions were actually used in the transportation study:

Q1. Do you use your automobile as part of your job? (1) no, (2) yes, 100 miles/day, (3) 75 miles/day, (4) 50 miles/day, (5) 25 miles/day.

Q2. Is the car you use for work provided by your employer? (1) no, (2) yes.

Q3. How far is your home from the nearest public transportation of any form? (1) 1 block or less, (2) up to 3 blocks ($\frac{1}{4}$ mile), (3) 6 blocks ($\frac{1}{2}$ mile), (4) 12 blocks (1 mile), (5) over 12 blocks (over 1 mile), (6) no public transportation available, (8) don't know, (9) no answer.

Q4. Have you used public transportation, if available, to commute to work from your residence? (1) no, (2) yes.

Q5. If you were to use public transportation, would you have to change routes or systems? (1) no, (2) yes, once, (3) yes, twice, (4) yes, three times, (5) yes, more than three times, (8) not familiar with the routes, (9) does not apply.

Q6. Would you consider car-pooling if available? (1) no, (2) yes, (3) don't know, (4) currently pool, (9) does not apply.

Q7. Would you consider van-pooling if available? (1) no, (2) yes, (3) don't know, (4) currently pool, (9) does not apply.

It has been said that Americans are in love with automobiles and that only extreme circumstances will force them to change.

Q8. At what price per gallon of gasoline would you consider pooling? (1) $1.50, (2) $2.00, (3) $2.50, (4) $3.00, (5) over $3.00, (6) never, (7) currently pool, (8) don't know, (9) does not apply.

Q9. At what price per gallon would you consider public transportation worthwhile? (1) $1.50, (2) $2.00, (3) $2.50, (4) $3.00, (5) over $3.00, (6) never, (7) currently pool, (8) don't know, (9) does not apply.

In the next questions, we would like you to rate the relative importance of certain aspects of driving yourself to work.

Q10. Rate the importance of the convenience of driving yourself to work as follows: (1) very important, (2) moderately important, (3) slightly important, (4) don't know, (9) does not apply.

Q11. Rate the importance you place on the safety of driving yourself to work as follows: (1) very important, (2) moderately important, (3) slightly important, (4) don't know, (9) does not apply.

Q12. Rate the importance you place on the privacy of driving yourself to work as follows: (1) very important, (2) moderately important, (3) slightly important, (4) don't know, (9) does not apply.

Q13. Rate the importance you place on the general tranquility of driving yourself to work as follows: (1) very important, (2) moderately important, (3) slightly important, (4) don't know, (9) does not apply.

Q14. If the government set a limit of 20 gallons of gas per week for the car you drive to work, could you get by? (1) yes, with no trouble, (2) yes, just barely, (3) no, (4) don't know, (9) does not apply.

The responses to the questions are listed in Appendix A and should be added to the cards or the file prepared for previous exercises in this section. Suppose that, as a researcher, you suspect that social factors play a large role in determining an individual's choice of public or of private transportation. Remember that regression is a technique fairly often chosen for this type of work. In it, the researcher suspects that certain variables, either singly or in combination, will predict a dependent variable, such as the ratio of public transportation use to private transportation use. This means that some questions predict the dependent measure, while others do not. Use program P1R to obtain a regression of all 14 questions on the dependent measure.

As a second problem, obtain a correlation matrix for the independent variables and a scatter plot of the observed predicted values for Q12. Request any other options appropriate to your understanding of the statistics involved. To simplify, use PUBTRANH as the dependent variable. You might check to see if the same questions predict the use of public transportation by the wife in the household. Remember that if you are using a BMDP file, CASES, FORMAT, and the /VARIABLE paragraph are not specified. Use of the DATA sentence is recommended for all analyses.

Answer:

```
/PROBLEM TITLE = 'PUBLIC VS PRIVATE TRANSPORTATION STUDY'.
/INPUT VARIABLES = 29.
FORMAT = '(F3.0,7X,F2.0,1X,F1.0,1X,F1.0,F2.0,3X,F1.0,1X,4(F2.0,1X)5(F5.0,1X)3X,
6(F1.0,1X)/10X,8(F1.0,1X))'.
CASES = 5.
/VARIABLE NAMES = ID,AGE,SEX,SES,VOTEDIST,NUMAUTO,MILESHUS,MILESWIF,PUBTRANH,
PUBTRANW,MIAUTOH,MIAUTOW,MIAUTO3,MIAUTO4,MIAUTO5,Q1,Q2,Q3,Q4,Q5,Q6,Q7,Q8,Q9,
```

```
Q10,Q11,Q12,Q13,Q14.
USE = PUBTRANH, 16 TO 29.
BLANK = MISSING.
MISSING = (16)0,0,0,0,0,0,0,0,0,0,0,0,0,0.
/REGRESS DEPENDENT = PUBTRANH.
TITLE = 'PUB VS PRIV TRANS ALL QUESTIONS ON HUSBAND'.
/PRINT DATA.
CORRELATION.
/PLOT VARIABLE = Q12.
/END
```

At this point, this answer should not be surprising. The USE sentence is important in limiting the number of variables available, especially when the independent
variables are not specified in the /REGRESS paragraph. In this case, an INDEPEN-
DENT sentence was not necessary because the assumption of all variables is correct.

When doing several regressions in one run, note that some of the parameters
cannot be set or reset for each analysis. For example, if you want to change the Y
intercept, then the entire program must be rerun.

Unlike this fairly simple example, regression problems frequently
involve large data bases, with hundreds of questions and thousands of
respondents. It is unusual to completely examine all the data in one
run; multiple runs are common in regression. This type of analysis,
then, is best done using BMDP files rather than cards. The savings in
time and money can be substantial. Go to the trouble of establishing
files suitable for your situation. Such files may be BMDP system files
or operating system files. Further savings can be obtained by retaining
the covariance matrix from the first run and using it in subsequent
analyses. (See the BMDP manual for details.) Note that using terminal
files instead of cards will also result in savings.

======== EXERCISE 8 ========

In this last exercise of the series, we will continue to work with the insomnia-reduction data to finally obtain an answer to the question of whether or not the experimental treatment reduces insomnia. There are three basic ways to analyze the data.
First, we could use the pre-treatment information as a covariate and perform a
one-way ANOVA between the groups on post-test. Second, we could treat the
design as a two-way ANOVA with repeated measures (pre-treatment and post-
treatment would be the first and the second measures, respectively). Or third, we
could use a one-way ANOVA of difference scores, i.e., the variable named CHANGE

that we used in previous exercises. Without debating the statistical issues, the first procedure is technically the proper analysis. To use this procedure with BMDP, we need to use program P1V. However, many readers are not prepared for covariance analysis. The second analysis procedure calls for program P2V and an understanding of two-way design. Therefore, in this exercise, we will use the third method (difference scores), keeping in mind that the mathematics of such scores is uncertain.

Using P1D, write the PCL for the insomnia-reduction study as a one-way ANOVA between treatment groups, with CHANGE as the dependent measure. You may use either cards or files, but a more meaningful analysis will be obtained if you use the entire set of 30 subjects from Appendix A. Ignore all options that are beyond your present statistical skill, but do include MAX, MIN, MEAN, and CORRELATION with the /PRINT paragraph.

Answer:

```
/PROBLEM TITLE = 'INSOMNIA REDUCTION'.
/INPUT VARIABLES = 7.
FORMAT = '(1A2,1X,2(F1.0,1X)2(F2.0,1X)2(F3.1,1X))'.
CASES = 30.
/VARIABLE NAMES = ID,TREATGRP,SEX,AGE,INSOMTYP,PREHOURS,PSTHOURS,CHANGE.
ADD = 1.
GROUPING = TREATGRP.
/GROUP CODES(2) = 1,2,3.
NAMES = TREATMNT,APLACEBO,CONTROL.
/TRANSFORM CHANGE = PSTHOURS - PREHOURS.
/DESIGN DEPENDENT = CHANGE.
/TITLE = 'ONEWAY ANOVA ON PRE-POST CHANGE NO COVARIATES PLOTS OR TRENDS'.
/PRINT MINIMUM.
MAXIMUM.
MEAN.
CORRELATION.
/END
```

There are few complications in this answer if one refrains from using options that are not understood in statistical terms. It is not recommended that you take the option of DEPENDENT = ALL. In this exercise, several one-way ANOVA's would result (one for SEX, one for AGE, one for INSOMTYP, and so on). This is particularly important, since only TREATGRP constitutes a true experimental variable, which can be manipulated and assigned by the experimenter. The others are organismic variables, which are properties of the subject. Organismic variables are usually assumed to be natural covariates of the experimental variable.

Note that covariates are considered to be independent variables by P1V. The experimental variable is considered to be a grouping variable. Thus, in our example, we grouped on TREATGRP (a true experimental independent variable), we defined CHANGE as the dependent measure, and we did not designate any covariates. It is good procedure to use the /GROUP paragraph to name the levels of the experimental independent variable. This paragraph is mandatory if there are more than ten levels or if cutpoints are used to determine levels.

The /PRINT paragraph is very straightforward, but by all means take the various options in the order in which they are listed. Orders that are not standard work properly until BMDP detects an error and then attempts to indicate its location. Remember that error messages are general rather than specific.

9.4   Pro Tips and Techniques for BMDP

In many ways, this section will benefit the experienced BMDP user the most. It also has great potential for causing misunderstanding and argument. There are innumerable tricks and flashy things that can be done with BMDP, and only a portion of them are known to this author. To further complicate the situation, each installation (computer center) is different enough to invalidate some of the suggestions that follow.

For these reasons, all of the following recommendations must be pretested if there is any doubt as to how or if to use them. When possible, a terminal should be used, since checking is so much easier. Also, consultants do not know all there is to know about BMDP. Check things for yourself; you may find an approach that has been overlooked. The *BMDP Newsletter* is a good source for innovative analyses not covered in the manual.

Remember that each version of BMDP is a separate entity, and what works for BMDP-75 might not work for BMDP-81, for example. Much the same is true for the variations among "host" computers.

Each of the topics discussed in this section is separate; none is dependent on what precedes it. Since the order of topics is one of convenience, browsing for areas of interest is perhaps the best strategy.

9.4.1   Formats

The FORTRAN formats that may be used are not limited to those discussed in the BMDP manual or to those mentioned in Chapter 6 (that is, F, I, X, and A). A common and useful additional format is the E, or exponential, format, which has the form E $w.d$. The width of the field is w, and d is the number of digits to the right of the decimal. The data may be entered in the E form also. Thus, this sentence:

```
FORMAT = '(1E8.2)'.
```

defines a field width of eight characters including two decimal places. For example, 74318E+3 is the equivalent of 743.18 $\times$ 10^3, or 743180. In effect, this is the everyday scientific notation that expresses numbers as powers of 10. This format is quite useful if very large or very small numbers must be input. There are two cautions when using the E format. First, precision in calculation does not automatically increase. And second, there is a limit on the exponent that will vary among machines (see your consultant). Note that many BMDP programs use E formats on output.

Another acceptable format is the double-precision format, or the D format. This is not very useful to the BMDP user because, although the data are twice as precise at input, the calculations are still in single precision. Some BMDP programs use double precision for specific calculations but not at the user's option. Structurally, D and E formats are the same.

FORTRAN programmers may wonder about Z and G format codes. The Z format is reserved for hexadecimal data—an unlikely need for social scientists even though it may function properly. G formats are identical to D, E, and F formats except that the form of the input data determines the format (D, E, or F), thereby allowing the programmer to avoid specifying the type of data. Again, not much need for this option exists. The general recommendation is to avoid both G and Z formats.

Another option that is functional in many BMDP installations is the scale factor, or P code. The scale factor is always a change in the number of powers of 10, i.e., the number multiplied by 10 to the scale factor. Thus, if 784.66 were output with an E10.3 format, it would be printed as 0.784E 03. If the format were 1PE10.3 instead, the output would be 7.847E 02. The P format may be used in conjunction with the D, E, F, or G format. Use caution with this format, since it functions in opposite directions for input and output.

A useful format for certain test development and test scoring situations is the L, or logical code, format.

There are two means of inserting comments into the data: the H format and the literal. Once inserted, these comments will then appear in any listing of the data. A code of 17H states that the next seventeen characters are to be treated as literal data, and as such they will not be analyzed. Alternatively, the literal can be enclosed in apostrophes, and then whatever has been enclosed can be inserted into the data. Note that in either variation the literal does not appear in the data but appears as part of the controlling FORMAT statement. For example, we could have

```
FORMAT = '(6F2.0,3X,10(F1.0,1X)(22Hdata collected 10may80)).
```

or

```
FORMAT = '(6F2.0,3X,10(F1.0,1X)'data collected 10may80')'.
```

The final format of interest is the T, or tabular specification, format. In effect, the T format allows tabbing to a specific column. T12 tabs to column 12, skipping columns 1 through 11. Unlike the tabulating on a typewriter, T-format tabbing can be performed in either direction, allowing rearrangement of the input data. The NAMES sentence must correspond to the tabbing order, or else the variables will be misnamed, for example:

```
FORMAT = '(6F1.0,T75,F5.0,T40,F6.0)'.
```

instructs BMDP to find six one-digit variables in terminal positions (or card columns) 1 through 6, to skip to column 75 and find one number of five digits, and to return to column 40 and find one six-digit variable. In many BMDP situations, the T format is a convenient means of ordering variables. And, of course, many programs do not have internal reordering provisions. This option also offers a way to avoid counting the number of blanks or ignored data. Simply tab to the beginning column of relevant variables. This process tends to reduce input errors.

9.4.2   Abbreviations and Plurals

Throughout this chapter, all instruction words have been spelled out in full. That is, the instruction word VARIABLE always works, whereas the instruction words VAR and VARIAB may or may not work, depending on which version of BMDP is being used. Though these differences are well documented in BMDP, they are not available in a convenient form. Accordingly, Table 9.4 is a compilation of the most troublesome abbreviations.

It may seem silly to the advanced user to discuss this issue of plurals, but experience demonstrates that all of us occasionally make a slip. In general, avoid the use of plurals. The specific effects depend on the exact details of each program as well as the nature of the local operating system. The only statement that consistently allows a plural is CASES. The supposition is that CASES was the original statement and CASE is an abbreviation. It is not worth the effort to determine whether or not other plurals also function correctly—just avoid them. Making a plural of an abbreviation also does not work, as in VARS.

9.4.3   Order of PCL (BMDP versus BMD) and Program Cycling

The sentence order within a paragraph and the paragraph order are at the user's discretion except for /PROBLEM and /END. However, the older BMD package does not group cases according to a specific variable but rather by counting cases. For example, if there are 15 cases

TABLE 9.4. Commonly Confused Abbreviations

Instruction	BMDP	BMDP-77, -79	BMDP-81
/PROBLEM	PROB	PROB	PROB
TITLE	none	none	none
/INPUT	none	none	none
VARIABLE	VARIAB	VAR	VAR
FORMAT	none	FORM	FORM
CONTENT	none	CONT, CNT	CONT, CNT
/VARIABLE	/VARIAB	VAR, VR	VAR, VR
MAXIMUM	MAX	MAX, MX	MAX, MX
MINIMUM	MIN	MIN, MN	MIN, MN
MISSING	MISS	MISS, MSS	MISS, MSS
/TRANSFORM	/TRANSF	/TRAN, /TRN	/TRAN, /TRN
GROUP (in /VARIABLE paragraph)	GROUP	****	****
GROUPING (in /VARIABLE paragraph)	****	GROUP	GROUP
GROUP (in /TEST paragraph)	none	none	none
STATISTICS	STAT	STAT	STAT
/STATISTICS	/STAT	/STAT	/STAT
CHISQUARE	CHI	CHISQ	CHISQ
WILCOXON (in /PROCEDURE paragraph)	WILCOX	WILC	****
WILCOXON (in /TEST paragraph)	****	****	WILC

****indicates instruction not available

per treatment or condition, then BMD automatically defines 16 to 30 and 31 to 45 as the second and third groups, respectively. Hence, shuffled data records are a disaster in BMD. Program control language that is out of order will produce erroneous results in a similar manner. Be cautious when switching from BMDP to BMD. Note that BMDP program P8V is structured as a BMD program.

Most programs in this series allow the sequential analysis of several problems without resubmitting the program. For example, in program P3D, the test paragraph may be repeated before the /END to produce separate analyses. This cycling does not require that the data be re-entered or placed in a BMDP file. However, all data used in any cycle must be read in at the beginning, so there is no opportunity for additional input.

9.4.4   Data Transformations and Conversions

The BMDP manual gives extensive coverage to the topic of data trans-formations and conversions, and it should be your primary source of information. Remember that all FORTRAN manipulations of the data must be written as subroutines, not as main programs. This means that transferred data must be in a form that can be passed from the sub-routine to BMDP. Usually, it is better to create new variables than to substitute a transformed value, particularly if files are used. Both resub-mission of the raw data and reconversion are irritating and tiresome tasks. Do not fail to discuss the use of FORTRAN subroutines with your consultant.

There are many useful transformations possible without resorting to subroutines. For example, all of the library functions (sines, cosines, tangents, etc.) and conditional logic (IF $X > 70$, then $Y = N$) are avail-able. They are written as part of the /TRANSFORMATION paragraph. Assuming that you are familiar with the appropriate chapter in the BMDP manual, we give the following examples of typical transforma-tions in psychology. Assume that the variable X represents the time required to complete a puzzle solution. We want to analyze the square root of X. For versions earlier than BMDP-81, the instructions would be as follows:

```
/TRANSFORM NEW = X.
NEW = SQRT(NEW).
```

If the square root of $X + 1$ had been required, we could have written:

```
/TRANSFORM NEWX = X.
NEWX = NEWX + 1..
NEWX = SQRT(NEWX).
```

For BMDP-81, we would have this:

```
/TRANSFORM NEWX = SQRT(X + 1.).
```

Note that only one operation per sentence is allowed (in versions before BMDP-81), and two periods are required when adding numeric con-stants (i.e., the value added is not the integer 1 but the real number 1.00). The arctangent, natural log, or log base 10 would all be obtained in the same way. Consult any FORTRAN text for a complete listing of these library functions.

If there are 15 variables to be transformed, a separate sentence or combination of sentences must be constructed. At some point, it be-comes easier to write a FORTRAN subroutine than to write individual transformation sentences. It depends on three points: the level of proficiency with FORTRAN, the number of variables, and the complexity of transformations. In many situations, it is simpler (though more laborious) to write a large number of simple TRANSFORM sentences. Remember, a subroutine, like any other program, must be checked to make sure that it functions correctly. One solution is to write the subroutine as part of a dummy FORTRAN program. If it works correctly with the dummy program, it will work with BMDP.

Transformations may be used as a means of selecting cases. This can be done with the USE, OMIT, and DELETE statements as discussed in the BMDP manual and in Sections 9.3.2 and 9.4.5, or it can be done by creating a decision network with LE, EQ, GE, etc. Experience indicates that for most purposes the BMDP sentence procedure is best for beginners.

The last transformation topic is the conversion of calendar dates to meaningful data. The usual practice of recording a date (be it birthdate, test date, or injection date) as month/day/year or even in military style as 10MAY81 makes the date unavailable for use as data. Think about it: an animal born on 020179 (February 1, 1979) appears younger than an animal born in December of that year (120779). There are, of course, many other things wrong with using dates just as they are written. It helps to revise the order to year/month/day, but this still does not create a continuous variable. The conversion routine below converts dates to consecutive days within the years 1979 and 1980. YR is the year variable, MO is the month, and DAYS is obviously days. The following routine could be used to determine a variable, such as the number of elapsed days after exposure until the first symptom of a reaction appears. For BMDP-81, we have

```
IF YR = 79. THEN YRDAYS = 00.0.
IF YR = 80. THEN YRDAYS = 365.0.
IF MO = 01. THEN MODAYS = 00.0.
IF MO = 02. THEN MODAYS = 31.0.
IF MO = 03. THEN MODAYS = 59.0.
IF MO = 04. THEN MODAYS = 90.0.
IF MO = 05. THEN MODAYS = 120.0.
(and so on through MO = 12.)
TOTALDAY = YRDAY + MODAYS + DAYS.
```

Those of you who are using older versions of BMDP would need to stretch the process:

```
TEMP = YR EQ 79.
YRDAYS = 00.0 IF TEMPYR.
TEMPYR = YR EQ 80.
YRDAY = 365 IF TEMPYR.
TEMPMO = MO EQ 01.
MODAYS = 00.0 IF TEMPMO.
```

```
TEMPMO = MO EQ 02.
MODAYS = 31 IF TEMPMO.
TEMPMO = MO EQ 03.
MODAYS = 59 IF TEMPMO.
TEMPMO = MO EQ 04.
MODAYS = 90 IF TEMPMO.
TEMPMO = MO EQ 05.
MODAYS = 120 IF TEMPMO.
(and so on through MO EQ 12)
TOTALDAY = YRDAYS + MODAYS.
TOTALDAY = TOTALDAY + DAYS.
```

Similar routines can be used to determine age from birthdate or age at testing from birthdate and test date. The point is that ages in this form possess the characteristics of true interval data, and we are thus able to avoid the illogical outcomes of working with dates in conventional form.

BMDP-81 offers yet another useful transformation option; we can change alphabetic characters to numeric with the CHAR function. In a study where the gender of the subject is coded M or F, conversion could be accomplished as follows:

```
IF (SEX EQ CHAR(M)) THEN SEX = 1.
```

This instruction would select males (M), since when SEX = F the variable SEX is set to zero (false) and is excluded because it is seen as a missing value.

Case selection may also be accomplished by the USE sentence. To select only those cases whose last name is JOHNSON, we would write

```
IF (NAME1 EQ CHAR (ƀJOH) AND NAME2 EQ CHAR (NSON)) THEN USE = 1.
```

Note that BMDP-81 uses character blocks in the CHAR function. In the above case, we have blocks of four. (See the discussion of A formats.) Presumably, the CHAR function could also test blocks of 1, 2, and 3. The important thing is to recall the nature of blanks in A formats; i.e., ƀJOHNSON and JOHNSONƀ are not the same.

9.4.5　Data Selection and Checking

The ability to sort (order) a data file (tape, disk, terminal lines, or cards) is a powerful technique that has been unavailable to BMDP users as an internal function prior to the 1981 version. SORT is an option in the newest version and is part of the /INPUT paragraph. The form is

```
/INPUT.
SORT = (insert variable subscripts or names).
ORDER = A.
```

Here ORDER selects an ascending (A) sequence. The code B specifies a descending sequence. Remember that the first variable will vary the slowest and the last variable will vary the most rapidly.

Sorted files are useful in a variety of ways, not the least of which is obtaining well-structured data listings. If you are using a BMDP release that does not have this feature, there are several alternatives. The simplest of these is to use a card sorter with a card file, but this can be a tiresome and brain-numbing task if there are more than a few variables to be sorted. A better alternative is to use the utility programs that are available at most computer centers. Especially useful are those programs that do a lot of traditional data processing. For example, IBM facilities often have a program named SORT as part of a library of utilities. To use this program, one must create a new non-BMDP file from the data or else make an external copy of a BMDP file and sort it. Remember that BMDP files include lots of headings, variable names, etc. Your local consultant should be able to help you. As a last resort, you might try to create a FORTRAN subroutine, but this is a substantial task.

The companion to sorting is merging, and it is available only through the use of FORTRAN subroutines. The BMDP-81 manual gives several examples. Merging is essential in studies where the data arrive over a period of time—for example, in a developmental psychology study in which children are tested once a year. All that was just stated about sorting is also true for merging. It can also be done external to BMDP. This is becoming a very straightforward process on interactive terminal systems.

Two final points need to be made on sorting and merging. BMDP now promises the capability to save a non-BMDP file (FORMAT = BINARY in the SAVE program). This would allow processing by a wide variety of programs, including SPSS and SAS. The second point is that BMDP can write files directly readable by SAS and P-STAT, and, conversely, BMDP can read files written by them. Sort and merge functions are standard parts of SAS; hence, one could write several files to SAS, use SAS to merge them, and then use SAS to write to BMDP. Although this is not as complicated as it sounds, it would most likely require the help of a good consultant.

FORTRAN Subroutines Almost anything that can be programmed in FORTRAN can be incorporated into BMDP as a subroutine. The only constraint is the arrangement of a BMDP file. There is a special programmer's guide to BMDP that gives complete descriptions of the files. The critical factor here is the time and effort needed to develop and test such procedures. Once you are familiar with the structure of BMDP files, there is no reason to limit yourself to FORTRAN, but it is up to you to make things compatible.

The USE Statement   When working with BMDP files or larger data bases, we find that the production of reams of unwanted information is a troublesome problem. By all means, limit the analysis with the USE sentence (perhaps in conjunction with the OMIT and DELETE functions as well). Although this hardly seems like an advanced-user tip, the problem is all too common. Remember, the development pattern of all packaged programs is to provide more and more service. It would be cynical to say that this means more and more arcane instructions, but the amount of material produced by a BMDP (or other) analysis is impressive. Make sure you know what you want and what it means, and then eliminate the rest. Recall that USE in the /VARIABLE paragraph refers to variable selection, whereas USE in the /TRANSFORM paragraph selects cases.

Missing Values   The XMIS function allows the user to assign a missing value to a variable. If several values are assumed missing, a transformation will be required (see page 53 of the manual of BMDP-81). Whether the blank or the zero will be allowed as missing values will depend on which version of BMDP and which host computer is being used. There are far too many combinations to list here, so ask your consultant about them.

Dirty Tricks with BMDP   The major analysis systems are actually corporate rivals, and as such they do not talk much about each other, except disparagingly. The point is that what cannot be done with BMDP can often be done with SPSS or with SAS. As discussed earlier, the files used are incompatible; but if you place SAS in the middle, you have an unbeatable system. If you have access to an interactive terminal system that allows private files (library), place your raw data in these files as card images. Then you can submit the files to any analysis package, since they all read cards. This avoids the problems of file compatability at a small sacrifice in speed. A binary file is another possibility, as all of these packages read binary files. A word of caution—remember that computers and installations vary just enough to complicate the transporting of files. Within a given system, however, the routine availability of private mass storage makes analysis by different packages a reality, whether the packages are compatible or not.

Interactive BMDP-81   Interactive statistical analysis is available in a limited form in BMDP-81. Certain programs are written so that they can dialogue with the user. That is, the program will request information in the form of input from a terminal. This is usually referred to as prompting. At the moment, this ability is not well defined and is not available at most installations.

9.4.6 Error Messages and Deliberate Errors

Messages to the user about errors can be divided into two classes: messages from the operating system and messages from the program itself. There are normally hundreds of system codes, and every system is different. Rather than list all of these codes in a bleak and partially superfluous appendix, Table 9.5 provides a list of the most commonly encountered operating system errors associated with using BMDP.

Error messages from BMDP itself have been discussed throughout this chapter. There are several key points to remember:

1. All such messages are rather general and give only hints.
2. The message system is dependent upon the expected order of instructions. Occasionally, errors may be pinpointed by rearranging the sentences and resubmitting them.
3. The error system does not attempt to cover all possible errors.
4. There are situations (see below) in which programs will function correctly despite error messages. Check these against known results before trusting them.

There is at least one BMDP situation that results in both an error message and a correct analysis. This is P2V as a one-way ANOVA with a repeated measure. This program is not specifically designed with one-way analysis in mind, and so it will always search for at least one more factor. The mean, standard deviation, and N will of course be zero. The result is division by zero, which in turn produces numbers infinitely large. The "overflow" and "underflow" messages will be printed, but the ANOVA table will be correct. There may be other situations of this type in BMDP, but you should always check against known results before deciding that an error is inconsequential.

9.5 Pro Tips and Techniques for Specific Programs

The descriptions of the programs that follow are not presented as the final word. There is always a variety of procedures that are proper for a given analysis rather than one ideal form. The suggestions are the result of experience, and they are certainly not a replacement for the BMDP manual. Suggestions are given only about the more popular programs. Programs of interest to the reader should be reviewed.

TABLE 9.5. Common Error Messages of IBM Systems

Error code	Message	Comment
ILF 010I	size (number is too large)	SCL error likely
ILF 013I	syntax error	SCL error likely
ILF 211I	invalid format code	happens often
ILF 212I	formatted line exceeds buffer length	for example, a 91-character line or card was specified
ILF 213I	reading past end of file	did not see PCL, or else data were found at end of file while looking for more data or for PCL (check data; this is a very common error)
ILF 219I	end-of-file read	
ILF 223I	invalid character in input record	
ILF 225I	data error; OLD PSW printed, followed by a string of digits, the eighth of which means:	
	9 = fixed point divide exception,	
	B = decimal divide exception,	
	C = exponent over-flow exception,	result of calculation is too large
	D = exponent under-flow exception,	result of calculation is too small
	F = floating point divide exception	
ILF 229I	missing "define file" statement	SCL or file PCL statements in error
ILF 236I	insufficient storage	consultant can help define greater storage
ILF 237I	record not found	program was looking for data not found

Adapted by permission from *IBM System/360* FORTRAN IV *Programmer's Guide,* Appendix A. © 1980 by International Business Machines Corp.

9.5.1   The P4D, P1D, P2D, and P3D Programs

P4D is the program to start with because it is very useful for verifying and proofing data. As discussed previously, the only quirks of P4D are the 80A1 format and the fact that BMDP files cannot be saved. When inputting from a terminal, it is prudent to place the information in a private file and copy to P4D for checking. Note that the PRINT options vary widely among versions.

The principal use of P1D is to establish the routine statistical properties of data sets. This program is useful for detecting "outlying" or wild cases. Detection of missing data is likewise very desirable. Distributions can be formed by grouping by select intervals for each variable (this is awkward but possible).

P2D continues the description of data but has additional features such as quartiles, skewness, range, and simple plots. Like P1D, it is very straightforward.

P3D is a convenient program for making one- and two-sample t-tests. The most frequent problems arise with the three separate grouping instructions, which are summarized here:

1. The GROUPING sentence in the /VARIABLE paragraph is the only mandatory instruction. The variable used for grouping may be a group code variable or a measured variable. If there are only two codes, or measured values (1, 0; yes, no; etc.), then the other grouping sentences are not needed. If, however, there are many group codes or if you wish to group on a variable such as IQ, the number of possible t-tests rises rapidly. The number of tests may be limited by using the /GROUP paragraph.

2. The second option, the /GROUP paragraph, limits the number of tests by defining intervals for continuous grouping variables. In BMDP, these are CUTPOINTS(X). IQ could be cut into blocks of ten, for example. By all means, label these intervals. The effect of limiting the number of tests, of course, is to limit the number of groups for comparison.

3. The third limiting procedure is part of the /TEST paragraph. To perform a test between students with a grade point average (GPA) of 2.00 and students with a GPA of 3.00, one states GROUPS = 2.00, 3.00. If CUTPOINTS are specified with a GROUPS sentence, then the groups are those implied by the /GROUP paragraph. If there is no /GROUP paragraph, then the reference is to rank order according to what was implied by the GROUPING sentence. In effect, this grouping option allows the user to further tailor the analysis. Its importance can be seen in the grade point average example, which would have 401 individual groups (0.00 through 4.00) if not restricted in some way. The /TEST paragraph also allows the user to limit the number of variables considered by P3D. It can be readily seen that since t-tests will be performed for each separate variable, it is important to be selective here, too.

The built-in time limit will prevent millions of tests, but thousands of meaningless tests could still be produced.

9.5.2   The P7D, P9D, P5D, and P6D Programs

Program P7D is clear and concise and requires only two comments. Make sure that the ANOVA instructions match your assumptions. The histograms often require some adjusting for maximum readability. Winsorization is a useful process, but P7D is a difficult place to learn what this process does and when to use it.

P9D allows you to define combinations of variables and then to determine frequencies, means, standard deviations, etc. The variable(s) of analysis is determined by a VARIABLE sentence in the /TABULATE paragraph. As mentioned in our discussion of P3D, use the /GROUP paragraph to limit the number of cells formed (combinations). Also, be careful about forming large numbers of zero cells. Remember that zero cells sometimes have meaning; e.g., there are no students with IQ's of 80 who have cumulative grade point averages of 4.0. On the other hand, sometimes zero cells are pure nonsense, as in the number of four-year-old college seniors.

Care in limiting the number of groups and variables is suggested with P5D and P6D. In addition to the ability to produce plots and histograms, these programs allow control of the dimensions of the charts. You can save paper by doubling up, or you can expand and contract them as needed. This means that the various diagrams can be matched to the size of a thesis page or other document. Do not forget, though, that manipulating the dimensions of a diagram can produce misleading interpretations.

9.5.3   The P4F Program (P1F, P2F, P3F, P8D, and PAM)

Note that in BMDP-81, P4F replaces P1F, P2F, P3F, P8D, and PAM. In P4F, there is commonly no /GROUP paragraph; instead, /CATEGORY is specified, and it functions in exactly the same way. Frequency tables are produced as sequentially paired (PAIR) combinations or as all combinations of row and column variables. It is highly recommended that you actually specify the variables of interest rather than search several hundred pages of tables for the one frequency table that is needed. Indeed, the PAIR default is apt to miss the needed variables because it matches row 1 with column 1, row 2 with column 2, etc.

All of this family of programs can produce a huge number of statistics, so do not ask for them all. Note that many of these statistics are produced in related clusters.

The P4F and P2F programs provide procedures for dealing with zero cells and with violations of strict independence assumptions.

The P4F and P3F programs provide tests of significance for multiway frequency tables, such as the log-linear model. P3F is almost incomprehensible unless one is familiar with these statistical methods. Thus, P3F does not have the self-teaching qualities of many other BMDP programs.

The P4F, P8D, and PAM programs provide a means of investigating and controlling the effects of missing data. P8D is particularly useful for examining large data bases whose characteristics are unknown, as when a colleague sends you the raw data of a survey and you do not know if there are any missing values or cases. PAM attempts to provide ways of adjusting for and replacing missing values.

9.5.4 The P1R, P2R, P9R, PAR, and PLR Programs

P1R and P2R are relatively simple programs if you understand regression. Beginners should first obtain advice about using them rather than simply letting BMDP determine the options. Regression programs can consume large amounts of computer time—time you may not be prepared to pay for. The plots produced can be adjusted for size, as was stated in the discussion of programs P5D and P6D. If you are using a very large data base and any regression program, try to determine time (cost) before actual analysis. See your consultant.

P9R, PAR, and PLR are complex regression programs, and they should be used only if the user has a clear understanding of the statistics involved.

9.5.5 The P1V and P2V Programs

The P1V program (one-way analysis of variance and covariance) is relatively simple except that the terminology used may be confusing for the first-time users. To a psychologist, an independent variable is the one that is manipulated or arranged by the experimenter. In P1V, however, these variables are referred to as GROUPING variables (as in P3D). You can group on the levels of a treatment (drug dosage, for example) or on the levels of an organismic variable, such as age or blood type. Confusion arises with the use of the term independent, since in P1V the INDEPENDENT sentence denotes the covariates (if any) and not the experimental variable. Covariates are not often of primary interest, but they are sometimes suspected of affecting the control variable and thus disguising it.

Problems may arise over the use of some similar terms: the GROUPING sentence of the /VARIABLE paragraph, the /GROUP paragraph, and the GROUP sentence of the /DESIGN paragraph. To clarify, we have the following explanation. GROUPING defines the variable of interest; /GROUP provides procedures for defining the legal codes for clustering, for naming, and for resetting the grouping variable; and the GROUP sentence provides a further means of restricting the groups compared and is also useful when several analyses are to be computed in one run.

When several dependent variables are specified in a DEPENDENT sentence, the result is separate analyses, as opposed to a multivariate analysis. The default is ALL, which means that if a dependent variable is not stated, there will be as many analyses as there are original variables less the variables listed in GROUPING and INDEPENDENT sentences. This is a costly procedure that often produces meaningless analyses (for example, an analysis of ID numbers).

There is no procedure for adjusting the size of the charts that PLOT creates. Note also that many of the options in the /PRINT paragraph apply to analysis of covariance.

P2V is a powerful program for analysis of variance and covariance and includes repeated measures. Remember, however, that it is strictly a fixed-effect model and that some of its advertised abilities (such as incomplete block designs) are possible only if one is prepared to specify the appropriate error terms. Also, multiple repeated measures must not be nested. Finally, P2V is not a multivariate program. If you specify multiple dependent variables, you will get separate analyses.

Even for experienced users, difficulty can arise with the /DESIGN paragraph. Ostensibly, this is where the exact analysis is defined. However, the terms used can create considerable confusion. For example, if a psychiatrist gives three different dosages of two drugs to a group of patients, it is traditional to refer to the dosages as levels and to the drugs as individual treatments. P2V uses the term LEVEL to refer to the number of observations within a repeated measure and not to subdivisions of a treatment. To make matters worse, repeated measures are called factors, and levels, as used above, are never mentioned.

The dosages used above are automatically determined by examination of each variable in the GROUPING sentence for subsets. Thus, we would specify this:

```
GROUPING = DRUG A,DRUG B.
```

The program would examine the codes contained in these variables to determine whether or not there are levels (subsets) of each drug. A number 1, 2, or 3 would appear in the data for each subject.

The term DEPENDENT refers to the dependent variable, as might be expected. In the psychiatric example above, it might be a measure of recovery. DEPENDENT also names the repeated measures (trial factors); in effect, each repeat is treated as a dependent variable. An important question is this: "How does the program distinguish between one analysis with ten repeated measures and ten analyses with no repeats?" This is where the LEVEL statement comes into play. It defines the number of repeats (levels) in each repeated measure (trial factor). In the psychiatric example above, if this:

```
DEPENDENT = DAY1,DAY30,DAY60,DAY90.
```

were specified without a LEVEL sentence, then four separate analyses would be performed. If the sentence LEVEL = 4 were included, then a single analysis with one repeated measure (of four observations) would be executed. LEVEL = 4,3 would define two repeated measures, the first with four repeats and the second with three.

Fortunately, the COVARIATES sentence means what it says. Remember that BMDP also refers to covariates as independent variables. A useful procedure is to perform the analysis both with and without covariates by repeating the /DESIGN paragraph.

Another troublesome point is the use of the subscripts instead of names in the design paragraph. It is especially difficult because GROUP, COVARIATES, and DEPENDENT sentences can refer to subscripts or to the variable names, while LEVEL is always a simple numeric designation. A subscript indicates the variable sequence position in the data. Remember that to BMDP all input is in terms of variables. Hence, GROUPING = 4,14 indicates two treatment factors (subscripts 4 and 14) whose levels or subdivisions are to be determined by examining the data. Experience has indicated that an instruction such as GROUP = DRUG A,DRUG B is less prone to error.

An alternative means of stating the model is to use the FORM sentence. It is the exact equivalent of the procedure outlined above, but is less ambiguous to many users. G refers to the grouping (treatment) relationships, Y to the dependent variable, X to covariates, and D to variables that are to be deleted from the analysis. For example, this sentence:

```
FORM = 'G,D,X,Y'.
```

states that the GROUPING factor is the first variable (subscript 1), that the second variable is to be ignored, that the third variable is a covariate, and that the fourth variable is the dependent variable. Of course, if the first four variables are the ID, name, address, and phone number, the analysis will be nonsense. Similar in logic to the FORTRAN FORMAT statements, each of the FORM designators may take multipliers and parentheses. For example:

```
FORM = '2G,2(DX)Y'.
```

translates to: there are two groups (variables 1 and 2), variable 3 is deleted, variable 4 is a covariate, variable 5 is deleted, variable 6 is a covariate, and variable 7 is the single dependent measure.

There are obviously a great many complications to the FORM sentence, and the BMDP manual has a thorough, if intimidating, discussion of these intricacies. The essential point to keep in mind is that this sentence is every bit as dependent on the subscript's variable order as is, among others, the GROUP sentence. The use of names instead of subscripts is recommended.

Determination of repeated measures and of individual dependent measures is accomplished through the position of the multiplier. 4(Y) denotes one measure repeated four times, and 2(3(Y)) denotes two measures, one repeated three times and the other twice. However, either 4Y or (4Y) indicates four separate dependent measures, and 2(3Y) specifies three separate repeated measure analyses, each with two repeats. Since this can easily lead to inappropriate FORM sentences, it is recommended that beginners expend the extra effort to do completely separate analyses of each dependent measure.

The NAMES sentence may be used with either /DESIGN paragraph and is quite useful for labeling the output from P2V.

A final point concerns the use of P2V with a one-way ANOVA. In a one-way design, there is no grouping (treatment) variable, which results in error messages in some versions of BMDP. The error stems from the multiplication and division by zero in the nonexistent cells. However, the ANOVA output is correct, and P2V will terminate normally.

9.5.6   The P3V, P8V, and P4V Programs

The P3V program is an interesting addition to the BMDP series (available only in BMDP-79 and BMDP-81). However, it is a program solely for the statistically sophisticated user. It does not perform a standard mixed-model ANOVA. Rather, it provides a maximum likelihood analysis. P3V is a good program to choose to learn this technique if you have a suitable text or an instructor who can provide direction. Simply put, this program is not self-teaching. For users who find P3V tempting, here are two suggestions. One, the /HYPOTHESIS paragraph refers to restricted maximum likelihood analysis and is not a part of standard maximum likelihood. Two, all variables in the FIXED instruction must have the levels defined in the /GROUP paragraph. Remember, this is an analysis procedure that is not universally accepted, but it will deal with unbalanced mixed models.

For many users, P8V is the most powerful of the ANOVA programs. It is available only in BMDP-79 and BMDP-81, but the same techniques are available as the 08V program with BMD. (See Appendix D for a discussion of BMD.) If you have successfully run the older version (08V), then P8V will seem relatively simple. Difficulty arises for those who have never used the older package because the structure of the P8V program does not follow that of the BMDP series as a whole. For example, there is no grouping paragraph or sentence; instead, factors (groupings) are determined empirically from a /DESIGN paragraph. This means that P8V does not require that codes for levels of a factor be included as a variable. Such codes, if present, must be ignored by the FORMAT statement.

A further constraint is that the order of the data must exactly follow the order implied by the /DESIGN. The order of processing is lexigraphical; i.e., the data are considered as a matrix that is read in the standard matrix order. Therefore, the LEVEL, NAMES, and MODEL sentences must all present the same order to P8V. The rightmost matrix index moves the most rapidly. For example:

```
NAMES = DRUG A,THERAPY,SUBJECTS.
```

requires that SUBJECTS change most rapidly, followed by THERAPY, and finally DRUG A. Note, then, that the data must be read in the order defined and (most critically) that P8V cannot recognize improper order.

In the older version (08V), assigning names was an option; it is a requirement in P8V. Each name must begin with a unique letter, since only the first letter is used in the resulting ANOVA tables. In both P8V and 08V, subjects (cases) are not implied but are explicitly stated as a factor.

Experience indicates that most errors are made in the MODEL sentence. It is easy to specify a model that cannot in fact exist. Table 9.6 will be of service in keeping the model relationship straight.

Since P8V will allow up to ten factors, there is the temptation to create very complex designs. Such designs are somewhat risky; significant ten-way interactions are difficult to interpret. Two other cautions are in order. P8V must have the same number of entries in every cell (this means equal subjects). Techniques for missing data should be applied before an analysis is done by P8V. Finally, there must be a one-to-

TABLE 9.6. BMDP P8V and BMD 08V Model Relationships

Example	Relationship	Comment
D,Q	crossed	symbols used are arbitrary
A,D,C,X	crossed	symbols used are arbitrary
A(D)	nested	A is nested in D
D,Q,A(D)	crossed and nested	A is nested in D but is crossed with Q
R,Q,S(RQ)	crossed and nested	S is nested in both R and Q
A,B,C,S(BC)	crossed and nested	S is nested in BC but is crossed with A
A,B,C,S(ABC)	crossed and nested	S is nested with all factors
A,B,C,S,S(ABC)	impossible	S could not be crossed and nested concurrently

one correspondence between the FORMAT statement and the variables named in the /INPUT paragraph.

Program P4V is not the place to begin the study of analysis of variance. It is a do-everything program that requires considerable skill on the user's part. One should begin with P1V or P2V. On the other hand, P4V is likely to become the most used analysis-of-variance program with experienced users. BMDP supplements its description with five separate technical reports indicating the flexibility of this program. It is worth the effort required to learn its use, but explaining it is far beyond the scope of this text. Use it, you will like it, but be prepared for a myriad of details.

9.5.7 Other Programs

BMDP also has the following programs: P3S, P1M, P2M, PKM, P3M, P4M, P6M, P8M, P9M, P3R, P4R, P5R, P6R, P7R, P1L, and P2L. P3S is a relatively simple program and generally provides no complications in its use. Of course, one should have a nonparametric situation before using this program. The cluster analysis, the factor analysis, and survival programs are techniques for the pros and require a thorough knowledge of statistics. Most of these techniques require that adjustments be made to the time defaults of the operating system. However, allowance for unlimited time could prove very costly. It is usually better to experiment with a small data set (or subset) in order to control time and cost. Use tape or disk files if available.

9.6   Summary and Comments on BMDP

Several years ago, there were clear differences between analysis packages. Today, however, the "big three" (BMDP, SPSS, and SAS) overlap extensively. An analysis-by-analysis comparison of the three packages seems pointless. Instead, we will review the virtues and faults of BMDP without comparison to the other packages.

An attractive virtue of BMDP is its segmented nature; each analysis is a separate program complete unto itself. This means that there is less demand on the host computer in terms of time and storage. This system is not bogged down with the complete set of programs just to do a simple t-test. On the other hand, this can be seen as a disadvantage in that the BMDP programs are not completely uniform. P8V is the most notable example of this lack of uniformity, but even these differences are minor and are easily overcome.

BMDP offers a wide selection of techniques for researchers using true experimental analyses. The factoring, clustering, and regression techniques are also extensive. However, it remains clear that BMDP is at its best with data sets that are comparatively restricted in size. This package can be used with the large survey data bases available, but it was not designed with this in mind.

Although BMDP is not an interactive package (does not dialogue with the user), it works well from a terminal, especially since the requirements for each program are summarized on a single page. Whether or not you will have to search through page after page of text to determine the next command is an important consideration because in many terminal systems you pay for each second of connection plus phone charges.

Appendixes

Appendix A. Extended Data Sets for Exercises

Data on the Conservatism/Liberalism Shift (see Table 7.3)

001	SMITH	291	01107	58570341373614	01979868778	976678765	421375421142	DMK
002	PARSON	301	02004	65551092363830	06887889667	878877767	319742558677	DMK
003	VALENTI	389	01029	45430722282910	03774879989	766488995	420644311132	DMK
004	WALSH	400	04100	52590171252909	02657445753	445573667	219112134213	DMK
005	JOHNSON	210	01035	60590171394006	01897665998	887859767	320132234121	DMK
006	LARSON	237	01029	45470742292512	02655676687	998799887	118566458567	DMK
007	JONES	311	01009	45470992383912	02698776699	996677896	118677547873	DMK
008	RALPHSON	342	04033	49480742344106	02778769768	888679677	319435743353	DMK
009	SCHWARTZ	334	04046	55560641252415	06676884565	845658676	422125731213	DMK
010	JOHNS	300	03335	60550202313308	01789979583	738895979	320334343432	DMK
011	WILSON	195	02060	57530752414103	06988999799	999799988	219667887654	DMK
012	VALESSON	105	03050	59581031374203	03977899799	979997789	118787787555	DMK
013	CLINT	325	02024	60500621352212	03879977777	777879976	421334211521	DMK
014	LUM	185	02029	47460721354206	03679977887	457886659	320454532353	DMK
015	STINSON	391	05041	45430391191815	02435446748	435558476	421456431243	DMK
016	WEINSTEI	275	05031	41430212293711	06576757878	787857666	218776557657	DMK
017	COOK	265	04004	48470211283807	02488945687	456874888	118488567894	DMK
018	MOORE	317	02099	51510372342505	02879995977	979987598	420324211571	DMK
019	FAMOUS	219	03037	50490652333908	01776686678	686678776	321867554535	DMK
020	WELK	275	05006	52470773393104	03998789969	789969998	119675867779	DMK
021	STARK	299	04014	46450821323709	03765957657	595765787	421454328222	DMK
022	FARB	314	03004	47470802393913	02897599999	985988579	320564333116	DMK
023	EDDISON	375	05004	50500911414203	06996579999	679889975	219678889755	DMK
024	LAP	210	05029	41400642343507	03657886568	886457678	219968886655	DMK
025	ANTONELI	215	03017	41391031392411	03997769999	999779999	118678878987	DMK

*Data from the Beer-Drinking Contest at Kelly's Bar
(see Table 6.8)*

```
0119831 37 27SEP81 43 17JAN82 40 02MAY82 C REILLY
0218837 11 23SEP81 23 18JAN82 31 29MAY82 S O'GRADY
0361821 41 26SEP81 42 16JAN82 40 09MAY82 S WILLIAMS
0420811 21 24SEP81 23 18JAN82 19 11MAY82 A AMATO
0520822 30 23SEP81 39 18JAN82 42 15MAY82 B MARKS
0619821 19 20SEP81 23 19JAN82 23 14MAY82 C THOMPSON
0718832 31 20SEP81 32 17JAN82 35 19MAY82 L TORONELLI
0817842 34 21SEP81 38 15JAN82 33 18MAY82 G DAY
0919821 29 27SEP81 30 18JAN82 30 07MAY82 F TETI
1019822 37 28SEP81 39 20JAN82 41 05MAY82 L SCHMIT
1121801 19 25SEP81 18 19JAN82 21 19MAY82 M GOLD
1122011 21 24SEP81 25 19JAN82 20 18MAY82 G MURPHY
1319822 44 20SEP81 41 18JAN82 47 05MAY82 D STEIN
1418832 41 25SEP81 46 21JAN82 50 17MAY82 H CALLISON
1518841 27 23SEP81 27 20JAN82 23 16MAY82 E KLING
1620812 39 24SEP81 40 19JAN82 40 15MAY82 T SMITH
1721822 41 22SEP81 45 16JAN82 44 14MAY82 S ERP
1820812 40 22SEP81 41 17JAN82 41 04MAY82 T MARTIN
1919821 30 25SEP81 30 19JAN82 31 02MAY82 K CLEMSON
2018842 21 20SEP81 20 20JAN82 39 02MAY82 F FOLSOM
2122812 23 26SEP81 25 21JAN82 41 19MAY82 N ANGELO
2223792 41 24SEP81 40 20JAN82 45 19MAY82 R FOX
```

Data from the Test-Anxiety
Survey (see Table 7.4)

001	20	62151014	031	31911	55	21
002	19	49040913	017	26111	42	29
003	19	72080012	019	38512	89	25
004	22	75076014	076	20512	75	28
005	18	51049813	085	32213	69	30
006	18	39041812	101	19813	94	26
007	21	31030714	064	28521	82	18
008	20	65067013	056	37021	79	15
009	19	49539013	034	36022	98	19
010	19	62035012	092	40022	71	14
011	21	62951014	091	26523	96	16
012	18	53048012	061	24323	65	17
013	20	60961013	020	30131	85	09
014	21	72051814	031	29531	86	03
015	21	56054013	041	26132	79	04
016	19	49058012	032	33432	91	08
017	20	60059014	005	32533	82	09
018	20	51053013	009	29033	90	07
019	19	60361022	092	31011	89	31
020	21	59071324	035	30911	75	35
021	22	31039013	021	19912	65	39
022	18	54062011	091	22512	80	09
023	18	69064021	095	31013	78	11
024	19	80053012	055	34013	95	03
025	20	72068024	077	31021	88	09
026	19	70555012	055	29921	75	15
027	18	70069021	017	34122	95	38
028	20	50060023	086	29322	91	32
029	21	62063014	017	32023	55	20
030	21	59060014	056	30023	59	38
031	19	63062012	055	31031	78	29
032	18	64048021	034	28531	85	29
033	18	32040511	020	20232	90	17
034	17	39062021	017	21032	50	14
035	19	61050022	003	23033	79	05
036	21	51053013	064	27533	85	04

Data from the Transportation Survey (see Table 9.2)

```
001 1    29 M 4 31    2 11 00 05 00 19000 09700 00000 00000 00000   8 6 3 3 4 5
001 2    4 3 7 2 7 8 3 3
002 1    41 M 2 22    2 27 09 00 10 25000 10000 00000 00000 00000   7 9 6 7 7 8
002 2    9 6 9 8 6 7 8 9
003 1    19 M 5 03    1 12 00 05 40 09000 00000 00000 00000 00000   2 4 1 3 5 6
003 2    2 3 2 4 1 1 7 1
004 1    24 M 4 17    2 07 09 25 00 06000 10000 00000 00000 00000   1 2 5 3 4 4
004 2    2 2 1 1 5 4 6 1
005 1    34 M 3 02    1 00 15 99 00 00000 14000 00000 00000 00000   2 2 3 2 4 4
005 2    1 1 2 3 6 2 1 2
006 1    28 M 3 17    2 15 17 30 00 11000 13000 00000 00000 00000   9 5 4 3 4 5
006 2    4 4 7 3 7 6 3 3
007 1    49 M 3 17    4 35 00 00 00 21000 09000 05000 11000 00000   1 2 1 3 4 4
007 2    1 3 4 5 4 2 1 1
008 1    61 M 1 13    4 06 00 00 00 17000 06000 01000 03000 00000   1 1 3 3 1 3
008 2    1 3 3 3 2 1 1 1
009 1    22 M 3 19    1 18 14 99 99 08000 00000 00000 00000 00000   7 9 3 3 4 5
009 2    4 3 7 2 7 8 3 3
010 1    29 M 2 13    2 41 09 00 15 15000 10000 00000 00000 00000   2 2 3 2 4 5
010 2    1 1 2 3 6 2 1 2
011 1    35 M 3 09    1 10 00 15 00 11000 00000 00000 00000 00000   3 3 4 3 5 5
011 2    2 2 2 3 7 3 2 3
012 1    48 M 4 09    5 11 12 00 00 15000 16000 11000 05000 07500   7 9 6 7 7 8
012 2    9 6 9 8 6 7 8 9
```

*Data from the Presidential Election
Voting-Behavior Survey (see Table 8.1)*

```
001 800929 19   M 24 9 9 9 9 9   9 07 20 17 00
002 800927 19   F 29 R R 9 9 9   9 07 09 00 14
003 800102 07   F 41 D D D D D   D 07 09 00 14
004 800930 19   F 37 R 0 D D 9   D 07 09 00 14
005 801004 08   M 36 R R R 9 9   R 07 20 17 00
006 801001 08   F 33 D D 0 9 9   D 07 20 17 00
007 801003 07   0 28 0 9 9 9 9   D 07 20 17 00
008 800929 19   M 31 0 D 9 9 9   D 07 20 17 00
009 800923 07   F 51 R D D D D   D 07 11 15 00
010 800915 07   M 59 R R D D D   R 07 14 00 09
011 800915 07   F 58 R R R R R   R 07 13 00 07
012 800915 07   M 64 R R D D D   D 07 13 00 07
013 800915 07   M 78 R R R R R   R 07 14 00 07
014 800928 19   M 45 R D R D D   D 07 20 18 00
015 800929 19   F 50 R R D D D   D 07 20 18 00
016 801006 08   F 39 R D D D 9   D 07 20 16 00
017 801006 08   M 49 R D D D D   D 07 21 02 00
018 801007 08   M 52 R R R R R   R 07 13 00 07
019 801007 08   M 29 R D 9 9 9   D 07 13 00 07
020 801003 05   M 44 D D D D D   D 07 13 04 00
021 801003 05   F 49 R D D D D   D 07 20 16 00
022 801004 05   F 58 R R D D D   R 07 20 16 00
023 801005 05   M 62 R D D D D   R 07 20 15 00
024 801005 05   F 60 D R D R D   D 07 20 15 00
```

Data from the Migraine-Headache Experiment (see Table 8.2)

0001	F	39	1	1	04	0820	04	0210	01	0150	3	4
0002	F	22	2	1	07	0530	00	0000	01	0050	5	2
0003	F	40	1	2	06	0400	06	0510	05	0400	7	8
0004	F	28	2	2	03	0700	03	0600	03	0300	5	4
0005	F	37	1	3	07	0210	05	0500	04	0430	6	3
0006	F	36	2	3	04	0630	03	0400	04	0400	7	7
0008	F	29	2	4	04	0430	05	0530	05	0530	4	6
0009	F	19	1	1	05	0530	03	0215	01	0200	2	3
0010	F	23	2	1	06	0700	01	0300	02	0230	4	4
0011	F	17	1	2	05	1030	06	0700	04	0800	7	8
0012	F	26	2	2	09	0915	08	0900	07	0900	5	6
0013	F	24	1	3	08	0830	06	0730	07	0600	7	8
0014	F	29	2	3	04	0730	04	0800	05	0800	6	5
0015	F	31	1	4	02	0230	02	0300	02	0300	4	6
0016	F	21	2	4	08	0530	07	0500	07	0600	8	7
0017	F	28	1	1	07	0600	02	0315	01	0200	6	3
0018	F	19	2	2	07	0430	07	0430	06	0400	8	9
0019	F	31	1	3	05	0430	06	0400	05	0500	7	6
0020	F	27	2	4	05	0730	04	0600	05	0600	6	6

Data from the Insomnia-Reduction Experiment (see Table 9.3)

01	1	1	22	07	050	075
02	1	2	17	01	043	081
03	1	2	18	03	038	045
04	2	1	19	03	044	043
05	2	2	19	07	051	055
06	2	2	18	04	030	038
07	3	1	21	01	041	071
08	3	1	20	08	055	065
09	3	2	20	10	025	045
10	1	1	17	03	035	060
11	1	2	18	04	050	070
12	1	1	21	07	044	049
13	2	2	22	08	052	053
14	2	1	18	01	060	059
15	2	2	19	03	035	040
16	3	1	20	03	039	038
17	3	2	21	04	044	043
18	3	1	20	01	049	050
19	1	2	20	01	054	065
20	1	1	19	06	049	064
21	2	1	20	05	048	049
22	2	1	18	03	047	048
23	3	2	18	07	065	066
24	3	2	19	06	041	042

Appendix B. Sources for Statistical Packages and Special Programs

BMDP and BMD: University of California Press
2223 Fulton St.
Berkeley, California   94720

SAS: SAS Institute Inc.
P.O. Box 10066
Raleigh, NC   27605

SPSS: McGraw-Hill Book Co.
Princeton Rd.
Hightstown, NJ   08520

SCSS: SPSS Inc.
Suite 3300
444 N. Michigan Ave.
Chicago, IL   60611

OSIRIS III: Institute for Social Research
University of Michigan
Ann Arbor, MI   48109

An excellent glossary of available packages and sources can be found in "Data Management and Statistical Analysis in Social Science Computing," by Ronald E. Anderson and Francis M. Sims, *American Behavioral Scientist,* 1977, **20**, pp. 367–409.

Numerous journals publish statistical program reviews. The following are particularly relevant:

American Statistician
Annals of Economic and Social Measurement
Behavior Research Methods and Instrumentation
Educational and Psychological Measurement
Journal of Marketing Research
Review of Public Data Use

Appendix C. OSIRIS III

OSIRIS is an integrated collection of computer programs for the management and analysis of social science data. It can perform many of the same analyses as the packages discussed in this text. However, OSIRIS performs the analyses in rather different ways. The central difference is in data management. OSIRIS is concerned with extensive data sets. In fact, the sets are so extensive that professional data processing techniques and procedures are required. OSIRIS is not oriented toward the type of data management common among social scientists who do not perform many surveys or who can simply toss a deck of data cards into a brief case.

The reasoning behind this orientation toward professional data management practices is clear once the sponsorship of OSIRIS is explained. The Center for Political Studies ISR, the Inter-University Consortium of Political Research, and the Survey Research Center ISR are the three principal sponsors of OSIRIS. (ISR is the Institute for Social Research at the University of Michigan, and the Consortium is an information intermediary that offers thousands of individual sets of data.) The data collected directly and indirectly by these organizations are all organized in OSIRIS-type files; hence, the OSIRIS system is a convenient tool for analyzing that data.

In many ways, OSIRIS is an admirable analysis package. It can do things that other systems cannot handle, but it is not a very easy package with which to begin learning about statistical packages. It is recommended that anyone new to computer-aided statistical analysis learn with SPSS, SAS, or BMDP. Facility with any one of these packages will be of great help in learning OSIRIS.

The OSIRIS-type files may be analyzed directly by SPSS and SAS. OSIRIS data sets have two parts—the data themselves and a dictionary that describes the data. This makes for efficient operation, but it means that special instructions are necessary when using these data sets with SAS or SPSS. There are restrictions on these cross-analyses that require careful reading of the manuals.

Although most of OSIRIS is beyond the scope of this text, one brief comment can be made to illustrate how fundamentally different it is. Many researchers collect data that require multiple cards (or lines) per case, subject, or observation. SPSS, SAS, BMDP, and this text all encourage the researcher to keep these cards grouped properly (for example, keep all of the responses to the first case together, etc.). OSIRIS requires that each card in a multicard data set be collected together with all other first cards, and so on. Thus, if it requires seven cards to describe your data, OSIRIS gives you seven separate data decks! Clearly, it is essential to have an ID number on every card and to have that number in the same position throughout.

With OSIRIS, the separate input decks will of necessity be merged into one, forcing the user to learn a program of little direct interest—the merge program. This situation is very characteristic of OSIRIS. In many ways, it is a package strictly for researchers with advanced computer skills.

Appendix D. BMD

BMD, or Biomedical Computer Programs, is the precursor of BMDP It has not passed into oblivion; in fact, some researchers prefer it to BMDP. There are several reasons for this. One reason is that many researchers learned with BMD and see no compelling reason to switch to the newer system. While BMDP-81 does have many extra features and a number of new statistical routines, it is not vastly different from BMD. In addition, some researchers prefer the style of instructions used in BMD. It uses a parameter-card form of instruction as opposed to the paragraph-and-sentence structure of BMDP-81. For example, both packages contain the 01D program for simple data description. BMD uses only three cards to give a complete set of instructions. This is characteristic throughout the package. BMD instructions are short and succinct, whereas BMDP specifications may be lengthy.

It is this compact form of instruction that led to the creation of BMDP because that form is somewhat difficult for beginners to learn. While not illogical or capricious, BMD is complex and nonintuitive. Where BMDP uses a declarative statement (CASES=150), BMD has the user place the number of cases in a certain pre-arranged card-column field, e.g., in columns 18, 19, and 20. But the field width is actually columns 15 through 20. Thus, it is a common error for beginners to place the number 150 anywhere in the field width, which could cause BMD to expect 150,000 cases instead of 150.

The more complex the statistics are, the more complex these dense instructions become and the more likely it is to make errors. In short, BMD is not an easy package for the beginner to use. Learn BMDP first; then the logic of BMD will merely be novel, rather than insurmountable.

Appendix E. SCSS

SCSS is the terminal-oriented version of SPSS that appeared in 1977. It is exclusively an interactive terminal program. Statistically, it is a subset of SPSS. The reason for the creation of SCSS was ostensibly to lessen the difficulties associated with learning a traditional package. The entire program is geared to a dialogue between it and the user. Whether or not SCSS lives up to this goal has not yet been determined.

The basic premise of SCSS is that the instructions for a given analysis are determined by a dialogue with the program itself. Three levels of conversation are possible: VERBOSE, NORMAL, and TERSE. These correspond nicely to this text's conception of the beginning, the intermediate, and the expert user, respectively. When you ask for a particular analysis, the type of response (also called a prompt) is set by these commands.

SCSS offers two further aids for the beginner—HELP and TUTORIAL. The HELP feature provides comments and directions for individual instructions. TUTORIAL is, in effect, a separate program designed to make SCSS self-teaching.

There are several very useful things to know about SCSS:

1. The program will make a number of assumptions that will surprise users of SPSS. These defaults are not incorrect. SPSS users would expect to state these instructions, but SCSS assumes them. These assumptions can be defeated by using the preempt command. One simply gives the preempt instruction, which does not allow SCSS to assume anything.
2. SCSS and SPSS files are not directly interchangeable. An SPSS system file is not readable by SCSS. However, SPSS can be instructed to create an SCSS file that can then be read by SCSS. In fact, many professionals prefer to create files with SPSS and then use SCSS to work with that data. It is clear that SCSS files are not easily converted to SPSS system files. Actually, SPSS can create both SPSS and SCSS files concurrently. As long as there are no restrictions on space, this procedure offers the best of both worlds.
3. Some SCSS prompts cannot be avoided. You must respond to the questions. A null (which is simply a carriage return) is a sufficient response.
4. As might be expected, SCSS offers a wide range of instructions for tailoring the output to a specific terminal. There is also a series of instructions that aids the user in remembering what files are in use and describes their characteristics.
5. Those familiar with OSIRIS files are aware of the concept of a "mask" for modifying data. SCSS provides a similar modification and labeling facility as well as the usual selection of file modification and labeling facilities. As usual, proper syntax is required.

SCSS is still evolving, and just where it is headed is uncertain. All things considered, though, it is reasonable to assume that it will change.

Appendix F. BCD to EBCDIC Conversion

EBCDIC character needed	EBCDIC punch code (IBM 029)	BCD punch required (IBM 026)	BCD multipunch
Letters A–Z			
Numbers 0–9		same	
&	12	+	
<	12-4-8	)	
%	0-4-8	(	
@	4-8	' (apostrophe)	
.	12-3-8	same	
-	11	same	
$	11-3-8	same	
*	11-4-8	same	
/	0-1	same	
,	0-3-8	same	
¢	12-2-8	none	B plus 8
(	12-5-8	none	E plus 8
+	12-6-8	none	F plus 8
|	12-7-8	none	G plus 8
!	11-2-8	none	K plus 8
)	11-5-8	none	N plus 8
;	11-6-8	none	O plus 8
-	11-7-8	none	P plus 8
None	0-2-8	none	S plus 8
_ (underscore)	0-5-8	none	V plus 8
>	0-6-8	none	W plus 8
?	0-7-8	none	X plus 8
:	2-8	none	Z plus 8
' (apostrophe)	5-8	none	5 plus 8
=	6-8	none	6 plus 8
"	7-8	none	7 plus 8

Note: Not all IBM keypunches use the 64-character EBCDIC system. Versions A and H produce only 48 characters and print different graphics (symbols) on the card. For example, the 12-4-8 code prints three different symbols [<, ⊐, and)] for Models EL, A, and H, respectively.

Appendix G. Data Sets for Student Solutions

Data Description for Prenatal Stress and Morphine Addiction Study

Variable	Number of digits	Number of decimal places	Function
1	2	0	case number
2	2	0	number of siblings
3	3	0	birth date
4	3	0	date at trial 1
5	3	2	weight at trial 1
6	3	2	milliliters morphine consumed at trial 1
7	3	0	date at trial 2
8	3	2	weight at trial 2
9	3	2	milliliters morphine consumed at trial 2
10	1	0	group number
11	2	0	ID number

Note: Group 1 = high stress
Group 2 = moderate stress
Group 3 = low stress
Group 4 = no stress
Dates: 2 digits for day, followed by 1 digit for month
Weights: in pounds
Morphine: 10% solution in water averaged over 10 days

Raw Data for Prenatal Stress and Mor-phine Addiction Study

1	3	031	044	2.01	2.45	076	2.29	2.37	1	03
2	3	072	065	1.94	0.51	038	2.03	0.61	1	33
3	4	011	024	1.75	4.12	076	1.70	4.04	1	15
4	4	013	305	1.22	0.38	297	1.95	0.35	1	56
5	4	291	304	1.50	0.51	306	1.91	0.48	1	43
6	4	131	293	0.95	5.09	275	1.59	4.61	1	20
7	4	022	015	0.87	3.47	037	1.41	3.99	1	25
8	4	081	124	0.94	2.34	116	1.52	2.89	1	25
9	4	132	135	1.37	0.21	147	2.00	0.43	1	31
10	4	081	103	1.54	4.01	135	2.31	3.87	1	07
11	5	041	093	1.92	3.45	085	1.93	3.54	1	18
12	5	072	065	0.89	0.49	067	1.91	0.41	1	37
13	7	262	255	2.21	0.39	277	2.10	0.04	1	49
14	8	091	034	0.91	3.06	056	1.49	3.11	1	16
15	8	272	305	1.61	0.44	297	1.89	0.48	1	59
16	4	011	024	1.98	3.61	035	2.04	3.53	2	01
17	4	082	105	1.73	0.48	097	1.99	0.41	2	03
18	4	311	015	2.19	0.37	027	2.08	0.51	2	55
19	4	061	044	2.01	4.02	066	2.10	3.89	2	10
20	5	022	055	1.89	0.45	037	2.00	0.41	2	32
21	5	191	214	1.75	2.33	206	1.84	3.14	2	17
22	5	041	044	2.19	4.03	076	2.10	4.01	2	08
23	5	022	054	2.14	3.45	036	2.02	3.99	2	35
24	5	172	165	1.85	0.53	187	1.99	0.31	2	46
25	5	091	104	2.22	3.41	116	2.31	3.99	2	12
26	6	033	026	1.98	0.29	048	2.01	0.39	2	58
27	7	271	284	2.13	1.37	266	2.05	1.41	2	29
28	7	141	164	1.91	3.45	166	1.89	3.63	2	19
29	7	021	054	1.89	4.25	076	2.02	4.01	2	06
30	9	202	255	2.19	0.49	212	2.10	0.91	2	40
31	4	091	114	2.05	3.94	117	2.07	3.69	3	11
32	5	041	054	2.11	4.01	046	2.10	3.89	3	34
33	5	261	294	1.99	0.44	286	1.93	0.48	3	23
34	5	291	304	2.16	1.55	017	2.01	1.90	3	28
35	6	052	075	1.85	0.52	067	2.07	0.48	3	38
36	6	023	046	1.78	0.41	048	2.05	0.29	3	51
37	6	111	134	1.82	3.39	156	1.99	4.11	3	09
38	6	033	056	2.10	0.29	048	2.00	0.24	3	53
39	7	141	134	2.00	4.31	126	2.01	4.09	3	13
40	8	271	284	1.85	3.52	276	1.97	3.62	3	22
41	8	052	075	2.14	1.31	067	2.10	1.00	3	42
42	8	112	105	2.10	0.59	127	1.95	0.29	3	48
43	9	012	065	1.91	0.43	077	1.98	0.36	3	30
44	10	271	264	1.87	2.59	286	2.09	2.94	3	47
45	10	042	055	2.01	0.61	067	2.09	0.48	3	52
46	3	272	295	1.93	0.40	307	2.02	0.39	4	44
47	4	051	064	1.75	3.51	046	1.99	3.99	4	02
48	4	191	194	2.09	4.25	216	2.20	4.31	4	21
49	4	062	075	2.11	0.39	067	2.03	0.30	4	39
50	5	023	046	1.92	0.24	028	2.04	0.21	4	60
51	7	061	072	2.04	4.99	056	2.01	4.95	4	14
52	7	072	055	1.84	0.90	067	2.08	0.52	4	24
53	7	013	046	2.31	0.32	038	2.19	0.24	4	45
54	8	021	034	1.94	5.43	046	2.12	5.01	4	04

Raw Data for Prenatal Stress and Mor-
phine Addiction Study (continued)

```
55  8 012 045 1.83 2.11 037 2.14 1.98 4 27
56  8 052 065 2.14 0.44 017 2.09 0.41 4 41
57  8 282 016 1.99 0.35 028 2.10 0.31 4 57
58  9 012 025 1.79 3.10 037 1.99 3.01 4 36
59  9 272 285 1.99 0.50 267 2.13 0.39 4 50
60 11 033 026 2.21 0.29 048 2.25 0.21 4 54
```

Data Description for Antiestablishment Attitudes in Off-Year Elections

Variables	Number of digits	Number of decimal places	Function
1	4	0	ID
2	3	0	gender (alphabetic)
3	2	0	age
4	2	0	education
5	4	0	state (alphabetic)
6	6	0	city (alphabetic)
7–32	2 each	0	questions Q1 to Q26
Second line or card:			
33	4	0	ID
34–37	2 each	0	questions Q27 to Q30
38	14	0	neighborhood (alphabetic)
39	1	0	race (alphabetic)
40	1	0	income

Raw Data for Antiestablishment Attitudes in Off-Year Elections

```
0101 fem 29 14 penn phil 0409140101091211070703090403030212110101070607050503
010104070309 fishtown      w 3
0101 mal 40 12 penn phil 0101070404030501060712121207070701070208030409050102
010203030301 queenvillage  w 4
0103 fem 18 12 penn phil 0301010405121201060712121207070701070208030409050102
010303090708 flourtown     b 3
0104 fem 51 16 penn phil 0406040309090802040501030108090707060707070708090304
010407040301 roxborough    w 5
0105 mal 16 10 penn phil 0302010606090103030201040707080104010304010601010404
010506040605 roxborough    b 6
0106 mal 41 14 penn phil 0911030708010804070407010308040708090102040707030405
010607070102 societyhill    w 8
0107 fem 35 12 penn phil 0604050109030401070404030409060605030103010605030701
010702030304 rittenhouse    w 8
0108 mal 37 15 penn phil 0404040803040707030903030704040701010304030202010607
010801020306 northeast     b 5
0109 mal 25 16 penn phil 0903070404030908032107010101010203070507090601070901
010901060503 northeast     w 6
1110 fem 26 18 penn pitt 0606090302090801031207010101010203070607090601070901
111006070406 west          w 5
1201 fem 24 11 penn pitt 0109070209070404070312111004212109090709110409070407
120107030301 north         b 4
1202 mal 29 12 ohio lima 0709010901030500040513111109111209120908120908070606
120202030901 none          b 3
1203 mal 30 21 wva  hunt 0605041101060603010404111209131211030911130708090509
120307010604 none          w 5
1204 fem 29 14 penn phil 0409140101091211070703090403030212110101070607050503
120404070309 fishtown      w 3
1205 mal 40 12 penn phil 0101070404030501060712121207070701070208030409050102
120503030301 queenvillage  w 4
1206 fem 18 12 penn phil 0301010405121201060712121207070701070208030409050102
120603090708 flourtown     b 3
1207 fem 51 16 penn phil 0406040309090802040501030108090707060707070708090304
120707040301 roxborough    w 5
1208 mal 16 10 penn phil 0302010606090103030201040707080104010304010601010404
120806040605 roxborough    b 6
1210 mal 41 14 penn phil 0911030708010804070407010308040708090102040707030405
121007070102 societyhill    w 8
1211 fem 35 12 penn phil 0604050109030401070404030409060605030103010605030701
121102030304 rittenhouse    w 8
1212 mal 37 15 penn phil 0504040805040707030903030804040701010304030202010607
121201020306 northeast     b 5
1213 mal 25 16 penn phil 0903070404030908032107010101010203070507090601070901
121301060503 northeast     w 6
1214 fem 26 18 penn pitt 0606090302090801031207010101010203070607090601070901
121406070406 west          w 5
1215 fem 24 11 penn pitt 0109070209070404070312111004212109090709110409070407
121507030301 north         b 4
1216 mal 29 12 ohio lima 0709010901030500040513111109111209120908120908070606
121602030901 none          b 3
1217 mal 30 21 wva  hunt 0605041101060603010404111209131211030911130708090509
121707010604 none          w 5
```

Troubleshooting Index

The following is a guide to the content of the exercises presented in the chapters on the three statistical packages.

Chapter 7   SPSS: Versions 6, 7, 8, and 9

Topics

Analysis of variance, 163–165
Creating new variables, 157–162
Cross tabulations, 152–154
Frequency analysis with continuous data, 151
Frequency analysis with integer data, 148
Integer versus continuous data, 149
Logical operators (GE, greater than or equal), 160
Multiple analyses, 155, 157, 159
Nonparametric statistics, 159–161
Regression, 162
Scattergrams, 156–159
Shortening instructions, 148, 157
Skipping variables with DATA LIST, 142
Temporary variables, 161, 164
t-Testing, 154–156

Instructions

DATA LIST, 140–144, 166
FILE NAME, 153, 155
FINISH, 153, 155
GET FILE, 157, 159

IF, 159–161
INPUT MEDIUM CARD, 148, 151, 153, 155
MISSING VALUES, 148, 151, 153, 155, 161, 164
N OF CASES, 148, 151, 153, 155
OPTION, 151, 153
READ INPUT DATA, 148, 151, 155, 165
RECODE, 161–163, 164–165
RUN NAME, 148, 151
SAVE FILE, 153, 155
STATISTICS, 153, 157
TO, 148, 157
VAR LABELS, 148, 151, 157, 159

Chapter 8   SAS: Versions 79 and 79.5

Topics

Analysis of variance, 224–225
Combining PROC steps, 229
Correlations, 214–215
Creating new variables (transformations), 214–217, 222–224, 232
Data arrays, 222–223
DO loops (new variables), 222, 231
File names (_DATA_, _NULL_), 215, 217
Frequency analysis, 207–210
Graphing (CHART instruction), 218–221
Logical expressions (LT, less than), 224
Multiple analyses, 211–212
Multiple versus compound PROC steps, 221
Passing variables between files, 215
Plotting, 220–221
Stepwise regression, 222–223
t-Testing, 216–218
Univariate statistics, 210–212

Instructions

BY, 214–215, 221
CHART, 218–221
DATA, 207, 210, 212, 214
DROP, 215, 217
FORMAT, 219
IF (modifying variables), 212, 214, 217, 222
INPUT, 207–210, 223–225
PROC, 207–214, 219
SET, 217
SORT, 214–215, 220
TABLES, 207, 209

Chapter 9   BMDP: Versions 75, 77, 79, and 81

Topics

Analysis of variance, 281–282
Chi-square analysis, 274–275
Continuous data (limiting of), 275
Creating new variables, 271–272, 276
Data lines (incorrect number of), 255
Formatting guidelines/problems, 256–257, 259–261
Frequency analysis, 274–275
Histograms and scatterplots, 276–277
Ignoring data, 270
Independent variables (more than 10 levels of), 275
Missing versus blank data, 266, 268–269, 271
Multiple analyses, 274, 277, 280
Pairwise versus crosswise tables, 275
Periods (omitting of), 251
P4D (special problems with), 264–265
Regression and correlations, 280
Saving data, 266, 269
Subdividing and labeling data, 266, 269–271
Subscripts, 273
Transformations, 271–272, 276
Variables, data, and names (relationships of), 253–254

Instructions

ADD, 271, 276
CATEGORY (GROUP), 271–272, 275
FORMAT=FREE, 258
GROUPING, 277
IS, ARE, and =, 250
MAXIMUM and MINIMUM, 266–268
PLOT, 277, 280
STATISTICS, 274–277
USE, 272, 280

General Index

Abbreviations:
  in BMDP, 284
  in SAS, 191, 226-228
  in SPSS, 167
ADD, 271
ALGOL, 17
Alphabetic information:
  in BMDP, 254, 257 (*see also*
    Chapter 6)
  in SAS, 195, 199, 200
  in SPSS, 128, 136, 141 (*see also*
    Chapter 6)
Analysis of variance (ANOVA):
  in BMDP, 280-282, 295-300
  in SAS, 223-224
  in SPSS, 163-164
APL, 17
ARE (IS and =), 250
Arrays, in SAS, 231-232
ASCII, 18
Assemblers, 17
ASSIGN MISSING, 172-173

Backspace (*see* Sections 4.2 and 5.3)
BASIC, 17
Baudot (baud), 18
Binary code, 18

Binary Coded Decimal (BCD), 14, 18,
  312
  to EBCDIC (*see* Appendix F)
Bit, 15
Blanks:
  in BMDP, 269
  in SAS, 191
  in SPSS, 126
BY, 153, 220
Byte, 15

Capacity designators, 16
Card (computer card), 14
  illustrated, 37
Card reader, 14
Carriage return (*see* Section 4.2)
Cases, number unknown:
  in BMDP, 256
  in SPSS, 177
Cassette (cartridge), 15
Category, in BMDP (*see* GROUPING)
CDC, 18
Central processing unit (CPU), 14, 18
Character constants (*see* Section
  8.6.1)
CHART, 218-220 (*see also*
  SCATTERGRAM and PLOT)

Chi-square analysis, in BMDP, 273-276
COBOL, 17
Codes, 17-18
Coding form (FORTRAN), 33
Commas:
  in BMDP, 257-258
  in SAS, 191
  in SPSS, 126
Compiler, 17
COMPUTE, 158, 171-176
Computer, structure of, 21 (*see also* Figure 2.1)
Continuations, of instructions:
  in BMDP, 252, 257
  in SAS, 191
  in SPSS, 129
Control fields, in SPSS, 126
Correlations:
  in SAS, 213-215, 239
  in SPSS, 135
CROSSTABS, 152-156
CRT, 18
Cursor, 49, 54, 55

Daisy wheel, 15
Data arrangement:
  column, 25-28
  lexigraphical, 25-28
  row-by-column, 25-28
Data bank, 19
Data base, 19
Data collection:
  correct procedures for, 31-33
  errors in, 30
  guidelines for, 23
DATA LIST, in SPSS, 124, 129-130, 140-144, 165-167
DATA statement, in SAS, 191-194 (*see* exercises in Chapter 8)
Data transformations:
  in BMDP, 271-275, 286-288
  in SAS, 212, 214-215, 219-223
  in SPSS, 175-177
Dates (*see* Transformations)

DEC, 19
Deck, 19
Disk (disc), 15
  in BMDP, 290 (*see also* SAVE)
  in SAS, 192, 248
  in SPSS, 131-132, 144, 157, 159
DO, in SAS, 222, 231-235
Dot matrix, 15
DROP, 206

EBCDIC, 18
  to BCD (*see* Appendix F)
EDIT, 146
END INPUT DATA, 177
ENTER (*see* Section 4.2)
Error messages:
  in BMDP, 291-292
  format of, 90
  in SAS, 235-236
  in SPSS (*see* Syntax)
ESC, 50
Executive, 17

Field widths, 90
File, 19
Files:
  in BMDP, 269
  in SAS, 193
  in SPSS, 131
FINISH:
  in BMDP, 255
  in SPSS, 132, 159
Floppy disk, 15
FORM, in BMDP, 297-298
Format(s) (*see also* Input):
  A formats, 101-103
  in BMDP (*see* Section 9.4.1)
  card, 90
  case selection, 110
  compound, 105
  decimal, 96-101
  errors in, 98
  F formats, 96-101
  floating point, 96-101
  guidelines for (*see* Tables 6.1-6.7 and Sections 7.2, 8.2, and 9.2)

hardware, 15
I formats (integer), 91–92
record lengths, 104–105
in SAS (*see* Section 8.6.6)
slash (record skipping), 109–116
special, in SPSS, 177
terms, 89
X formats, 91–92
FORMAT IS FREE, 258–260
FORMAT IS SLASH, 261
FORMAT IS STREAM, 260–261
Formatless input:
    in BMDP (*see* Sections 9.2.2, 9.2.3,
        9.2.4)
    in SAS (*see* Section 8.2.2)
    in SPSS (*see* FREE FIELD)
FORTRAN, 16, 87
    in BMDP, 250, 282–283, 286, 289
    in SPSS, 173
FREE FIELD, 137–140
Frequency analysis:
    in BMDP, 273–276
    in SAS, 206–210
    in SPSS, 146–152

GET FILE, 131
GROUPING, in BMDP, 293, 295, 297

Hardware (hardwired), 13
Hexadecimal code, 18

IF:
    in SAS, 203, 233 (*see* exercises in
        Chapter 8)
    in SPSS, 160–161, 171, 175
IF ... THEN, 215–218, 230
Implied lists:
    in BMDP, 268 (*see also* TO)
    in SAS, 209, 219
    in SPSS, 128, 142, 148
Input, 17 (*see also* Format(s),
        Formatless input)
    in BMDP, 256–262
    column, in SAS, 198–200
    guidelines, for BMDP, 257
    list, in SAS, 200–202

    in SAS, 192, 195–202
    in SPSS, 136–138
INPUT FORMAT, FIXED, 127–130
INPUT FORMAT, FREE FIELD,
    137–140
INPUT MEDIUM CARD, 128
Instruction sequence (*see* Syntax)
Integer data, 149, 153
Interpreters, 17
IS (ARE and =), 250

JCL, 19

Keypunch, 14 (*see also* Chapter 5)
    control switches, 61–62
    error correction with, 66–67
    keyboard, 62–68
    physical aspects of, 58–60
    program drum (memory), 68–74

Lexigraphical data arrangement, 25–28
Logical operators:
    in BMDP, 271
    in SAS, 234–235
    in SPSS, 159, 160, 170

Machine language, 17
Mark-sense forms, 33
MAXIMUM, 267–268
Memory, 14
Merge, 19
Microprocessor, 14
MINIMUM, 267–268
Missing values:
    in BMDP, 268–269, 290
    in SAS, 203–204
    in SPSS, 148, 151, 154–157, 162,
        167, 172–173
Modem, 15, 45, 47
Monitor, 17
Multiple analyses:
    in BMDP, 285
    in SAS, 204–206
    in SPSS, 155

NULL, 194, 215